PRACTICAL FLYING

A MANUAL ON FLYING DEALING
WITH THE MOST RECENTLY
ADOPTED METHODS OF TUITION IN
THE COMPLETE FLYING COURSE.
WRITTEN AND ILLUSTRATED BY
EXPERIENCED FLYING OFFICERS.

A NIGHT LANDING
Aided by the light of a parachute flare, dropped from the
machine, and wing tip flares.

PRACTICAL FLYING

*Complete Course of Flying
Instruction*

BY

FLIGHT-COMMANDER
W. G. McMINNIES, R.N.

WITH AN INTRODUCTION BY
MAJOR-GENERAL W. S. BRANCKER, C.M.G.
COMPTROLLER-GENERAL OF EQUIPMENT OF THE ROYAL AIR FORCE

ILLUSTRATED BY
FLIGHT-LIEUTENANT E. L. FORD, R.N.

WITH A CHAPTER ON THE MEDICAL ASPECTS OF AVIATION BY
H. GRAEM ANDERSON, M.B., Ch.B., F.R.C.S.

NEW YORK
GEORGE H. DORAN COMPANY

Printed in the United States of America

6466296

INTRODUCTION

THIS book appears at a propitious amount; the Royal Air Force is in the throes of creation, and the strong individualities of the Royal Naval Air Service and the Royal Flying Corps are in the course of fusion.

It is the work of a Royal Naval Air Service officer, and besides being a valuable addition to existing official publications, it reveals to the pilots of the Royal Flying Corps the principles and methods of the Service with which they are to share a great and glorious future.

Aviation has shown the world some wonderful steps in progress during this War, not the least noteworthy of which has been the advance made in our methods of training. Experience, courage and patience have built up a system which has completely revolutionised old ideas, and pilots to-day are performing evolutions with the utmost confidence which eighteen months ago meant almost certain death.

The responsibilities of the flying instructor are great, and his work hard and unceasing. His is the dull round of duty day after day, without the glamour and excitement of service in the Field, and often without the appreciation which it deserves. It is, however, on his courage, patience and energy that we depend for the maintenance in the future of that individual superiority which our pilots have so clearly demonstrated in the past.

Every day confirms the growing importance of aviation in war; indeed, it seems that all other means of bringing the enemy to his knees have almost ceased to progress, whilst our aerial fleets go on expanding and improving until at last they will bring us victory.

20th March, 1918.

W. Brancker
Major General

PREFACE

E VERY prospective pilot who is really interested in his work must have been struck with the dearth of literature dealing with practical flying. When the writer was learning to fly he searched everywhere for practical information on aviation, and studied all the available literature on the subject, but found most of it lacking in the very features he required. Even to-day, most books on flying are either far too advanced, dealing principally with the theoretical aspect of aviation, or else contain historical and personal, instead of practical, reminiscences. There seem to be few books written by experienced pilots in such simple language that a man who had never seen an aeroplane before might understand and appreciate the information, hints and advice given. It has been the writer's object to produce such a book, giving as much useful and practical information on elementary flying as space permitted. It is not pretended that it possesses any great literary merit, but it is hoped that it may be of use to those now learning to fly, as a good deal of the matter was originally given in lecture form to officers when undergoing their preliminary training at a flying school. It does not pretend to appeal to the expert pilot or the individual interested in the more technical or theoretical side of aviation. The idea has been rather to present to the prospective pilot, in terms that can be easily understood, a general impression of how flying is taught and learned. Forewarned is forearmed, and it is hoped that, by studying the advice and hints contained in the chapters on elementary flying, the beginner will be able to avoid many of the mistakes so commonly made while under instruction.

At the moment of going to press, the fusion of the Royal Naval Air Service and the Royal Flying Corps is taking place, and by the time the book is in the hands of the reader it is anticipated that the flying services will be consolidated under the title of the Royal Air Force.

Owing to the rapid changes which take place in flying, training systems and machines may also vary from time to time as new and improved methods are introduced, and, therefore, due allowance must be made for them in reading the book. If any important corrections are necessary between the time the bulk of the pages go to press and publication, they will be found in the addendum at the end.

Both R.F.C. and R.N.A.S. officers who were instructing at the same flying school as the writer have rendered him valuable assistance in producing the book, and his thanks are due to them for their suggestions and advice.

The sections of the Ordnance Survey maps and Characteristic Sheet, pages 121, 122 and 139, are reproduced with the sanction of H.M. Stationery Office.

CONTENTS

CONTENTS

CONTENTS xiii

CHAPTER XII

THE GROWTH OF CONFIDENCE—A WORD BOTH TO INSTRUCTORS AND
PUPILS

Instilling Confidence—After a Crash—Advantages of Considerable Practice—Advanced Dual Control—When a Pupil Can
Fly—The Time Taken in Learning 207

CHAPTER XIII

THE MEDICAL ASPECTS OF AVIATION WITH SOME NOTES ON SUITABLE
CLOTHING

By H. Graem Anderson, M.B., Ch.B., F.R.C.S., Temp.
Surgeon R.N., Attached to St. Mark's and the Belgrave
Hospitals

Physical Fitness—Reflex Actions—The Visual Reflex—The Audi-
tory Reflex—Tactile and Muscular Reflexes—The Balancing
Reflex—Drinking and Smoking—Over Confidence—What to
Wear—Frostbite—Air Sickness 214

APPENDIX

Flying Instruction Notes in Brief—General and Condensed
Hints in Definite Stages for the Guidance of Pupils . . . 219
Glossary of Terms Commonly Used in Aviation 229

PRACTICAL FLYING

CHAPTER I

Preliminary Considerations

BEFORE deciding to take up aviation, either as a hobby or as a profession, it would be as well for the aspirant to a pilot's certificate to consider whether he is suitable for it, both mentally and physically. Generally speaking, anyone of average health and intelligence can be taught to fly, but at the present time there are comparatively few who will make really first-class pilots. In peace time the demand will be greater for the ordinary class of pilot, and a much lower level of skill will be required than in war time.

Which Men Make Good Pilots?

It is not an easy matter to determine beforehand which men are likely to make good pilots and which are not. Generally speaking, the average man will make a fair pilot. There are a few people who are born aviators, and there are some, either owing to natural nervousness or because flying does not appeal to them, who will never become pilots at all.

Sometimes, before undergoing any instruction whatever, a pupil is subjected to fancy flying with a view to testing his nerves; but such a trial is apt to be misleading in its results. It is far better to instil confidence in a pupil by showing him how safe and secure he is in ordinary flying than to attempt to scare him in his first flights by steep spirals, nose dives, or loops. The growing confidence of a pupil in himself and in his machine should be cultivated.

Many people imagine that motoring or motor cycling forms an excellent apprenticeship to flying; but beyond the

1

fact that these pastimes provide the learner with useful mechanical knowledge, they will not be found to be of particular value to the aviator. It is the man who has been accustomed to riding and outdoor games who proves quickest at picking up the feel of an aeroplane and whose eye more rapidly adapts itself to the speeds and angles encountered when flying.

During his first few flights the pupil will have to accustom himself to the feeling of being in the air. This is quite one-third of the battle, and it is for this reason that observers and balloonists, who have already overcome the strange sensation of being up aloft, learn the actual handling of the machine more quickly than the complete novice.

When Not Suited to the Work

Under present conditions, it is a mere waste of time, both for the instructor and for the pupil, if the latter is not really keen on his work. If he decides, after five or six hours' dual-control instruction in the air, that he is not suited to the work, or has not sufficient stamina to carry on with it and overcome any preliminary nervousness that may assert itself, he had far better have left it alone in the first place.

It must not be thought from these remarks that aviation is a difficult or dangerous pursuit. Although great prominence is given in the daily Press to accidents in the air, these occur very seldom when the total number of pupils under training and the number of hours flown daily are taken into consideration. Accidents are generally due to inexperience on the part of pupils in their early training, which is probably speeded up, increasing the risk in consequence, more than it would be if the call for pilots were not so urgent. Apart from these causes, air accidents are very few and far between.

An aeroplane is a very simple type of machine compared with a motorcar, for instance. There are far fewer parts to go wrong. There is practically no transmission to give trouble, no gearbox, clutch or differential to break or require daily attention. The tyre problem on the aeroplane hardly exists. A dozen control wires, generally dupli-

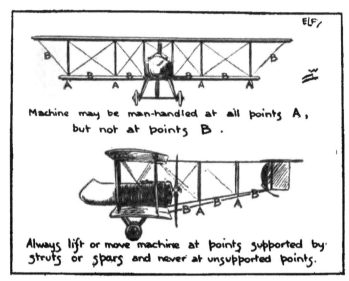

Ground training. Where and where not to man-handle an aeroplane. Always lift under struts and spars, and not under unsupported lengths. Never lift or push by holding on to the leading or trailing edge of a wing.

cated, are the only vital parts of the machine where failure might mean an accident. Compare them with the hundreds of parts in a motorcar which may go wrong, and some idea of the safety of the flying machine, from the mechanical point of view, can be obtained. Again, motorcar accidents are often due to the carelessness of the drivers themselves: collisions on blind corners or at crossroads, sideslips and skids on tramlines may be quoted as familiar examples. In the air there is unlimited space for machines. They are not confined to certain narrow tracks, and hence accidents are scarce where trained pilots on standard designs of machines are concerned.

Where accidents do occur generally is in getting off or in landing, but the risks attaching thereto can be practically eliminated by care on the part of the pilot. Modern flying—except that under war conditions—has arrived at a

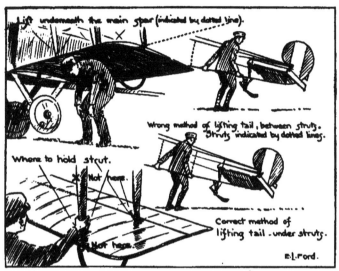

The proper way to lift the wings and tail of an aeroplane.

point where it has already been considered as a rival to other methods of travel, and it is probable that, in the near future, it will compete successfully with the motor-car or mail train over long distances, hence the wonderful opportunity for the younger generation to take up this new science and to be ready to make use of the opportunities it will offer when the war comes to an end.

Physical Requirements

To return to the case of the prospective pilot. It is generally admitted that he must be as physically fit as possible. His sight should be perfect, and although there are cases of pilots with half normal vision, necessitating the use of spectacles, there is the ever-present danger of the glasses breaking or becoming dimmed with oil or rain, and a consequent danger in landing.

Sound heart and lungs are the next most important points. A war pilot may be required to stay for several hours at heights varying from 15,000 ft. to 20,000 ft., and unless he is in the best possible health he will be unable to do this. Of course, under peace conditions the requirements would not be so strenuous, the average height for commercial flying probably being approximately 4000 ft. or 5000 ft.

A Preliminary Study of Aviation

All Service pilots have to pass a fairly stiff medical examination, so that there is not much chance of the physically unfit becoming aviators. Once through the medical examination, the prospective pilot is drafted away to a flying school, where, probably, he will have his first close view of an aeroplane. He may spend several weeks on the ground before making his initial dual-controlled ascent. This time can be employed to great advantage by a close study of the machine and by learning from the mistakes committed by pupils in a more advanced stage of training.

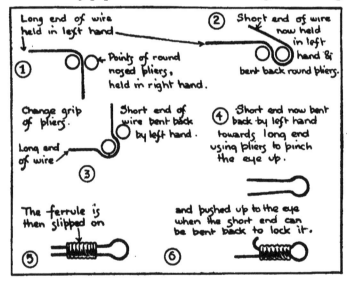

How to make loops in piano wire for attaching to the turn-buckles, etc. The top points of the round-nosed pliers used for this operation only are shown.

A keen pupil will spend as much time as possible in studying aviation handbooks and such literature dealing with the subject which may come his way from time to time. He should also do as much practical work on the machines as he can. It is always better for him to do the work himself than to watch it being done by others. For instance, a pupil will learn more in an afternoon by helping to dismantle a machine after a crash than he would have done by watching for several weeks the mechanics effecting minor adjustments to a machine in the shed.

Preparatory Ground Training

The following is a summary of the various points which should be studied in a preliminary ground training:—

(1) The pupil should first of all make certain that he is physically fit enough to fly, and that he is sufficiently keen and interested in aviation to be able to overcome any preliminary nervousness or difficulties that may be experienced in the early stages, otherwise he will be simply wasting his own and his instructor's time.

(2) During the pre-air work period of instruction the pupil should familiarise himself with the mechanical details of the machine on which he is going to be taught. He should then find out how the controls, levers, taps, switches and instruments work, and what is the proper speed to fly, glide and climb the machine and to run the engine.

(3) He should be able to do a certain amount of practical work on the machine, i.e., change plugs, start up the propeller, splice wire, make loops and eyes, etc. He should learn how to man-handle an aeroplane, at what points he can lift and move it, and those where it is undesirable to exert force.

(4) He should study as much aviation literature as he can obtain, and learn, at any rate, the rudiments of mechanical flight, the construction of aeroplanes, and the principles on which the petrol engine works. The last is most important.

CHAPTER II

The Theory and Practice of Flight

I T is necessary for the pupil to understand why an aeroplane flies, and thoroughly to master the objects and uses of the various controls.

Why an Aeroplane Flies

At first glance, it is not easy to see why an aeroplane flies. The pupil may have watched a machine run along the field for some distance and then gradually leave the ground. He has seen no lever move. Indeed, if he had been in the machine, it is doubtful if he would have detected any movement of the controls, and yet the machine, which was running along the ground a second or two before, is now in the air and, apparently, climbing quickly. An aeroplane flies because the propeller, driven by the engine, forces the planes, which are set at a small angle to the air, through the air at high speed, and a lifting action is thereby set up. Naturally, if the lifting action, depending on the speed at which the machine is forced through the air, is not great enough to lift the weight of the machine, the aeroplane will continue to run along the ground. Once sufficient speed has been attained, however, it will rise in the air, if the elevator is in the correct position.

It must always be remembered that it is the relative speed between the aeroplane and the air that assists the machine in flight. Ground speed is only identical with air speed when there is no wind; thus, if a machine were flying against a wind of 20 m.p.h., and the necessary difference of speed between the machine and air, in order to maintain the aeroplane in flight, were 50 m.p.h., the speed of the machine over the ground would only be 30 m.p.h. If, however,

7

the machine were to turn down wind, its ground speed would increase to 70 m.p.h. to reach its flying speed of 50 m.p.h., which always remains constant within certain limits. Many pupils find it very difficult to realise that their speed through the air is independent of the speed of the wind, although their speed over the ground is very greatly influenced by the speed of the wind, which may be adverse or favouring. For this reason it is possible for a machine to remain stationary over the ground if it be flying against a very strong wind, in which case the speed of the wind and the flying speed of the machine would be identical.

An Explanation of the Controls

The object now should be to discover how the machine is controlled. If the pupil inspects the pilot's seat, or, better still, sits in the machine on which he is shortly going to be taught to fly, he will find projecting from the floorboards between his knees a vertical control column, generally known as the "joy stick." His feet will be touching the rudder bar, and by working the stick and bar, and by observing the effect they have on the controlling surfaces of the machine, he can gain an excellent idea as to how the aeroplane can be made to climb or descend, turn, and fly in any desired direction.

The Rudder Bar

Consider the rudder bar first. This probably is a metal-lined piece of wood pivoted about its centre, so that the pilot's feet can swing it backwards and forwards round that centre. It is set athwartships, and at each end are control wires, sometimes duplicated, which connect the rudder bar with the ruddering surfaces of the machine. The rudder is situated at the stern, and consists of one or more vertical surfaces set parallel with the fore-and-aft line of the machine. Controlling wires are adjusted to such a length that when the rudder bar is dead athwartships the controlling surfaces are perfectly parallel with the fore-and-aft line. With the rudder bar square, the machine will fly straight;

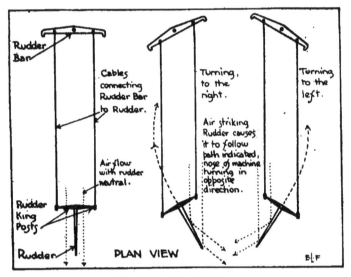

The effect of ruddering.

but when the right side is pushed forward it pulls the controlling surfaces to the right from its point of attachment (see sketch), and so opposes more surface to the wind on that side, which causes resistance and forces the tail of the machine round in the opposite direction, i.e., to the left. If the tail goes round to the left, the fore part of the machine will turn to the right, so that, to cause the machine to turn to the right, all that one does is to press the right foot forward on the rudder. The reverse is the case when it is desired to turn the machine to the left.

The "Joy Stick"

To make a correct turn, however, it is necessary to do more than press the rudder. The vertical control column, which has already been mentioned, is now called into operation, and must also be moved in the desired direction. This control column is pivoted about three-quarters of the way down and can be moved backwards, or forwards, or sideways from this point. It is used to control the elevator and ailerons of the machine.

The Elevator

The elevator, like the rudder, is situated at the back of the machine, except in certain old-fashioned types, where a front elevator is fitted. The elevator is a horizontal con-

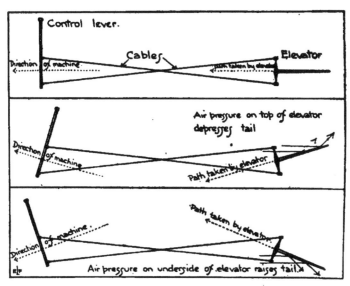

The elevator. The effect of the fore and aft movement of the control lever on the elevator and the machine as a whole. The control is pulled back to make the machine climb, and pushed forward to make it descend. The centre position the machine flies level when the engine is running.

trolling surface, or flap, set parallel with the main planes, and capable of being worked upwards or downwards from its hinged or pivoting forward point. It is connected to the "joy stick" on control column by wires (generally duplicated for safety's sake), the wires being arranged above and below the pivoting point of the control lever, so that when the lever is pushed forward the elevator surface moves down from its forward hinged end, presenting more surface to the air, causing a resistance which sends the tail of the machine up and the nose of the machine down. It will be seen from this that, to make the machine descend, the control lever must be pushed forward, and to make it climb it must be eased back. In the latter case the elevator surface will be raised above its neutral line, so that it presents its top surface to the air; the resistance encountered then sends the tail of the machine down and the nose up, and the machine is now ascending.

When the elevator is neutral, the control wires being connected to the joy stick, the latter must also be set in the neutral or vertical position. Where the control wires are connected directly between the control column and the elevator, it will be obvious that, to ensure the correct control, the wires must be crossed before being connected to the elevator, i.e., the wires from above the pivoting point of the control column lead to the underside of the elevator and vice versa. (See sketch.)

All the movements of the control lever are natural ones, i.e., to make the machine descend the lever is pushed forward; to climb, it is pulled backward; and to turn to the right or left, it is moved in those directions.

How "Stalling" is Caused

The attitude of the machine in the air is controlled by the elevator, as already explained. When the machine is set to fly at an increased angle the resistance of the wings is greater, the angle of attack having been increased so that the speed decreases and the machine climbs. When the machine is set by the elevator to fly downwards slightly, the resistance set up by the wings attacking the air is decreased and the machine increases its speed; so that, if a pilot knows at what air speed his machine flies level at a given number of engine revolutions per minute, he knows that any increase in the air speed must mean that he is flying downwards, and that any decrease indicates that he is climbing. If he allows the air speed of the machine to drop below the point where sufficient air is dealt with by the planes in a given time to maintain the machine in flight—i.e., if he allows the machine to stall—he will lose control, and only regain it when the air speed has again been increased by diving to a point when the air dealt with in a given time is sufficient to maintain the machine in flight. Before stalling, he will notice a general sloppiness in the controls, indicating the fact that an insufficient amount of air is passing the planes and their controlling surfaces in a given time to have the desired effect on the ailerons, elevator and rudder. He should then immediately put the nose of his machine down to increase the air speed and so regain control.

Banking and Turning

It is now necessary to explain that the control lever as well as the rudder bar must be moved in the proper direction in order to make a turn. If the rudder only were operated, the machine would be inclined to skid its corner, as it were, and, later on, the nose would go down. It would slip outwards on its turn and would lose speed. To make a gentle turn in the air the rudder must be applied, together with the correct amount of bank, in the following way :—To make a gentle right-hand turn the right rudder is pressed forward and the control lever is moved to the right at the same time ; to come off a gentle turn the control lever is moved across in the opposite direction, past the central position—that is to the left—and as the machine approaches the horizontal again a slight amount of left rudder may be given. Both controls are then centred. What happens when an acute turn is made will be discussed later in the chapter dealing with dual-control instruction.

The Ailerons

The pupil can now trace out the wires that connect the control column with the ailerons, or controlling flaps, fitted to each wing. In some machines there is an aileron, or controlling surface, to each of the four wings. In others, only

the top wings are so fitted. The principle on which the
ailerons operate is just the same as in the case of the other
controlling surfaces, i.e., the rudder or elevator. When
the control lever is moved to the right in order to make the
machine turn to the right—that is, the right-hand side of the
machine will be down and the left up—the controlling surface fitted to the left wing or wings, as the case may be, is
pulled down. The amount is very slight: the trailing edge
of the aileron may be, perhaps, half an inch below the trailing edge of the wing, but the additional resistance set up by
the extra surface exposed to the air causes that wing to
rise, while the reverse is the case with the right-hand wing,

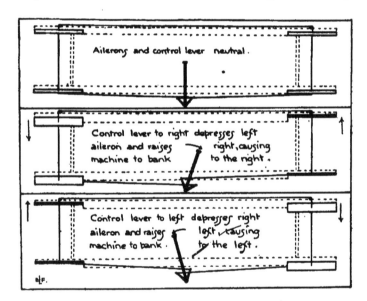

The ailerons. The sideways movement of the control lever when
it is pressed to the right makes the left aileron go down and the
right aileron come up, which causes the machine to fly right wing
down. To make the machine fly left wing down the control lever
is moved to the left. The sideways movement of the control lever
is used in banking the machine for turns and in correcting air
bumps, which cause a wing to rise or drop.

the aileron of which is connected by a balancing wire to the other aileron, and is arranged so that, when the latter is drawn down, it is slightly raised, thus helping the right-hand wing to become depressed owing to the added resistance on the top surface of its aileron, which is now raised slightly beyond its neutral position. On some machines the wings themselves are made to move or warp. The control is just the same as on the more modern type fitted with ailerons.

The Engine Control

Having mastered the details of the control of the machine, the pupil can next proceed to discover how the engine is controlled. If the engine is a stationary type, as opposed to the rotary, he will find his throttle lever and petrol tap not much different from those used on motorcars. He will also find the ignition switch, which should have its "off" and "on" positions clearly marked. In addition, there may be various pumps, the purpose of one of which will be to maintain air pressure in the petrol tank, where the fuel is not gravity fed. The pupil should find out which way all the taps and levers open and shut, and what are their functions and use. He should study the array of strange-looking clock-faced instruments and gauges which he will find on the instrument board in front of him as he sits in the pilot's seat. He will notice the air speed indicator, which gives him his speed through the air and not over the ground, except in cases where there is no wind; and also his height indicator, which may be graduated in hundreds and thousands of feet. The engine revolution counter may be driven direct off the engine shaft, in which case it will give him the number of revolutions per minute of the engine itself, or off the camshaft, in which case it gives only half the number of revolutions per minute performed by the engine. Petrol and oil gauges present no novelty to the man who has some motoring experience; but a compass and sideslip indicator may puzzle the pupil to begin with. When the instruments have been inspected and their operation explained and understood, the pupil should find out at what speeds, by instruments, the machine flies level, glides, and climbs; at what number of revolutions per minute it is safe to run the engine; how many gallons of petrol and oil the tanks hold; how much petrol and oil the machine consumes per hour, and, consequently, how long it can remain in the air without replenishment.

Study the Engine

Next he should turn his attention to the study of the engine. If it is one of the stationary type, it may follow motorcar practice. The cylinders may be set V fashion, with fours pairs of Vs in line, as on the Maurice Farman elementary training machine, or it may be a fixed engine, with the cylinders set radially round the crankshaft, as on

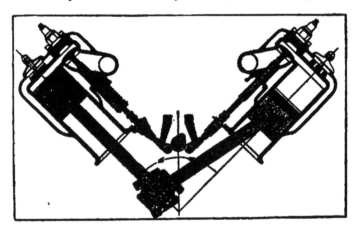

General arrangement of an eight-cylindered engine, showing two cylinders at 90 degrees, with central camshaft.

the Anzani-Caudron. If a pupil is being taught to fly on a more highly-developed and complicated engine, such as the Gnome, Clerget, or Le Rhone, he must study it with the greatest care, and make certain that he understands how to operate the throttle and fine adjustment levers without choking the engine. The stationary type of engine possesses ignition, carburetter and valve gear very similar to those employed in motorcar practice. If he does not possess any knowledge of the principle on which the petrol engine works, he had better at once obtain a reliable handbook on the subject, such as the "Motor Manual," and commence a study of it without delay, for the more he knows about his engine the more valuable will he be to himself and his comrades in after days.

Principles of Aeroplane Construction

A preliminary study of the machine should include a comprehension of the principles on which it is built and rigged. The pupil will find that the wings, which carry the weight of the machine, consist of main spars and ribs braced together by piano wire. Over the whole is stretched fabric, which is held in place on the ribs by stitching and copper rivets. The fabric is then doped by a special process to preserve it from the weather and to tighten it up. Between the wings, in the case of a biplane, are the inter-plane struts, set vertically. The whole is then braced together in cellule form by wires tautened by turnbuckles. The weight of the engine, passengers and body of the machine in the air is taken by what are called the flying wires, which run diagonally from the top plane inwards and downwards to the fuselage. The wires leading from the top plane outwards and downwards to the fuselage are called landing wires, because they take the weight of the machine when it is on the ground. In some machines, the upper and lower planes are assembled in one length, in which case the fuselage, or nacelle, is bolted to the centre section struts and rests on the centre of the lower plane. On others, the wings are divided into right and left-hand pairs—upper and lower —always taking "right" and "left" as referring to the machine when seen from the pilot's seat, and are bolted up to the fuselage and centre section. The fuselage of the machine is a long, thinly tapered and generally more or less rectangular box. At its forward end it accommodates the engine, whilst the lower wings are secured to it at each side by bolts passing through their front and rear spars. The passenger's seat is generally found immediately

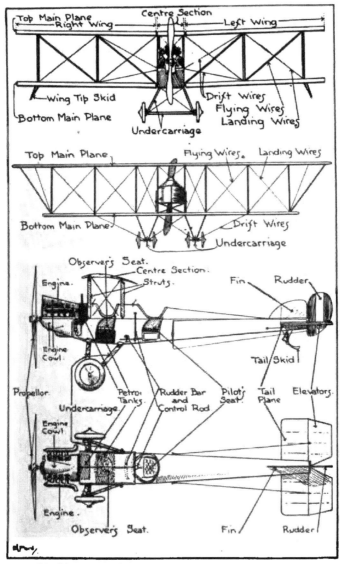

Top: **Machine front view, composed of right and left-hand planes bolted to the centre section. Bottom:** Top and bottom planes as one unit. A typical fuselage in elevation and plan view.

17

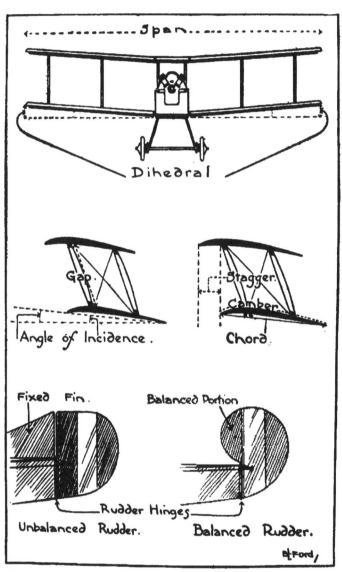

The three top sketches illustrate common aeroplane dimensions, and the lower two the difference between a balanced and unbalanced rudder. The pupil should familiarise himself with common aeroplane terms and their meaning as early as possible.

behind the engine, and behind it in turn comes the pilot's seat.

Arrangement of Tanks and Seats

The passenger's seat has four vertical struts rising from the fuselage round it, and above it is the centre section plane roughly corresponding with the width of the fuselage. To it are attached the upper planes, right and left. From this it will be seen that the engine is carried in front of the planes, the passenger's seat directly under the centre section of the top plane, and the pilot's seat slightly to the rear. The fuselage is tapered off to the rear, where the fixed tail plane, elevator flap, vertical stabilizing fin (if fitted) and rudder are accommodated. The petrol and oil tanks are found in the forward end of the fuselage, generally just behind and slightly above the engine, although, in some cases, a pressure-fed tank is fitted under the passenger's or pilot's seat. Generally speaking, however, the aim of the designer is to centre all the weight as much as possible, so as to secure ease of manœuvring. The whole of the weight of the machine on the ground is supported by the under chassis, or undercarriage, fitted under the fuselage and midway between the main planes. The undercarriage is attached to the wheels and axle of the machine, through the medium of some form of shock absorber, such as the Sandow elastic, so as to decrease the shock of landing, or "taxying" over rough ground, as much as possible. A tail skid fitted at the rear of the machine serves the same purpose in relation to the tail, which would otherwise drag along the ground until flying speed was obtained and might seriously strain both itself and the rear part of the fuselage.

Further practical knowledge may be gained by the pupil if he is able to spend some time in the erecting and engine shops. He will learn how to cover a plane with fabric, how to dope it, how to true up a machine that is nose or tail heavy, or flying with one wing down. Simple work, such as wire splicing, making loops and eyes in piano wires, and the correct method of fitting Sandow shock absorbers to the landing wheels, is all easily learnt, and will certainly be

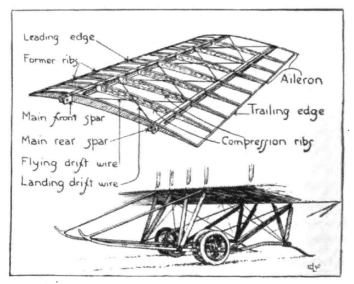

Leading edge
Former ribs
Aileron
Main front spar
Trailing edge
Main rear spar
Compression ribs
Flying drift wire
Landing drift wire

Principles of aeroplane construction. Constructional details of an aeroplane. Top: What a wing looks like when the fabric is removed. Bottom: A typical undercarriage with double pairs of landing wheels and arrangement of skids.

of great value to the potential aviator at some period of his career. At the same time it is quite possible to become a first-rate pilot without a great deal of technical knowledge, just as many people can drive cars or motorcycles well without knowing much about their working. Other things being equal, however, the man who knows most about the practical and mechanical side of his machine and engine will be more valuable than the man who knows little or nothing.

CHAPTER III

Elementary Principles of Aeroplane Engines

IN school flying, one of the principal troubles to be guarded against is engine failure. Not that this is so frequent as it was in the early days, for aeroplane engines are very much more reliable now and are very much better understood, but the necessity for studying the engine and its peculiarities and of obtaining a good working knowledge of it cannot be too strongly impressed upon the pupil. It may save quite a lot of trouble and very possibly some forced landings.

While learning to fly, the pupil should take a personal interest in the engines and machines he has to handle, and if at any time any trouble is experienced with an engine similar to his he should endeavour to find out the reason and the remedy. He should acquire such practical knowledge as whether it is safe or advisable to continue flying with one or two cylinders misfiring; what black smoke belching out of the exhaust pipe signifies; why certain machines are difficult to start up in cold weather, and how this failing can be remedied; how much petrol and oil the tanks hold, and how much fuel the machine uses per hour; and the causes and cures for the hundred and one little troubles that will crop up from time to time during his course at the flying school.

The Four-stroke Principle

Practically all aeroplane engines, as in motorcar practice, work on what is known as the four-stroke principle. Any handbook on the petrol engine, such as the "Motor Manual," explains this detail, but a rough outline of the working of a four-cycle engine may be given. Imagine a cast-iron jar turned upside down with a close-fitting, but shorter, iron jar, also inverted free to slide inside it. Fixed to the inside of the inner jar is a rod, which is enlarged and hollowed out at its lower end so as to fit over a crank. Carrying the simile still farther, at the top of the larger jar are two doors which can be opened and closed by a suitable mechanism at the desired moment, from which lead two pipes, one to the carburetter or device in which the petrol and air are mixed in the desired proportions so as to form an explosive mixture, and the other to the silencer, or exhaust expansion chamber, from which the burnt gases can issue into the air. The door, or valve, to which the carburetter pipe leads, is called the inlet valve, and the door, or valve, opening into the exhaust pipe, is called the exhaust valve.

Engine Parts

There is also a machine called the magneto, capable of supplying an electric spark at the correct moment in the space between the top of the shorter jar and the under side of the top of the largest jar. This space is generally known as the combustion chamber, while the larger jar represents the cylinder and the inner and smaller one the piston. The rod from the piston to the driving shaft is called the connecting rod, and the shaft is, of course, the crankshaft, to one end of which the propeller itself can be fitted. Both the valves are opened at the correct moment by a system of gear wheels, driven off the crankshaft and cams, which lift the tappet rods and rocker arms in contact with the valve stems, but they are closed by springs, called the exhaust valve spring, in the case of the exhaust valve, and the inlet valve spring, in the case of the inlet valve. The magneto is also driven by the engine, and sends an electric current along a rubber insulated wire to the sparking plug, which is screwed into the combustion chamber with its sparking points inside, and its insulated terminal, to which is attached the cable from the magneto, on the outside of the cylinder.

The cycle of operations is divided into four strokes of the piston. The first is the suction stroke, with a downward movement of the piston; the second is the compression stroke, with an upward movement of the piston; the third is the explosion or power stroke, with a downward movement; and the fourth is the exhaust stroke, with an upward movement of the piston again. After the four strokes have taken place, the series starts again, and as each up and down stroke of the piston represents one revolution of the crankshaft, which may be doing from 1200 to 2000 revolutions per minute, it can easily be understood that the strokes fol-

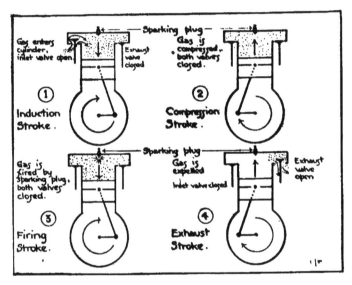

Diagrammatical illustration of the principle of the four-stroke engine, showing the cycle of operation.

low each other with extraordinary rapidity. It will also be noted that only one stroke in four is a power stroke, and therefore the engine has only one power stroke to every two revolutions of the crankshaft. Thus, if the engine were running at 1200 revs. per minute (usually expressed as r.p.m.), it would mean that only 600 power strokes, or explosions, were taking place every minute.

Having grasped the cycle of operations, the pupil must

next familiarize himself with the working of an engine in greater detail. The piston in the course of its movements up and down inside the cylinder never reaches the closed top of the cylinder; there is always a small space, called the combustion space, left between the top of the piston at the upward limit of its travel and the under side of the top of the cylinder.

How Power is Developed

When the piston descends on the first stroke the inlet valve opens and the explosive mixture is sucked into the combustion chamber from the carburetter by way of the inlet pipe by the displacement caused by the downward travel of the piston. The inlet valve is then closed by the action of its spring, and as the exhaust valve remains closed during the first three strokes of the cycle of operations, the combustion chamber is sealed gas tight. The piston then ascends and compresses the mixture already received into one-fourth or one-fifth of the space it originally occupied. When the piston is at the top of the second (compression) stroke the magneto is so timed that it sends an electric current along the rubber insulated wire to the sparking plug points inside the cylinder, when the spark takes place and fires or explodes the compressed mixture. This explosion drives the piston down on its third, or power, stroke and imparts energy via the connecting rod to the crankshaft of the engine. On the last stroke of the cycle of four the exhaust valve opens and the piston in its upward course drives the burnt and used gases before it, past the exhaust valve into the expansion chamber and exhaust pipe. When the piston has reached the top of its travel it has driven out most of the used gases. The exhaust valve closes and the cycle of operations commences again, with a downward travel of the piston and the opening of the inlet valve. In passing it may be noted that a gas-tight joint is made between the piston and the inner walls of the cylinder by means of piston rings, which fit in grooves in the piston and tend to spring out against the walls of the cylinder.

Valve Systems of Aviation Engines

This cycle of operations is common to the great majority of aviation school engines, among them being the Renault eight-cylinder air-cooled, the Curtiss, and the R.A.F. A slightly different system is employed in some rotary engines, notably the Gnome and the Monosoupape.

In the Gnome engine the mixture is taken into the crankcase, and enters the combustion chamber by means of a valve in the head of the piston. This valve is worked automatically, and so does not need a system of timing gears, cams, tappets and rocker arms to operate it. The valve is caused to open by the suction in the combustion chamber on the intake stroke, and a spring returns it to its seating when the piston ascends again to compress the mixture. The exhaust valve is operated in the ordinary manner and is carried in the head of the cylinder. See sketch of this valve arrangement on page 28.

The Monosoupape, or single valve, engine works on the four-stroke principle, but only employs one valve, which is carried in the cylinder head as in the Gnome. This valve is open to the air, and on the inlet stroke allows air only to enter the cylinder. It closes, however, some time before the piston reaches the bottom of its travel, or stroke, so that suction is set up inside the cylinder while the piston is still descending. There are passages connecting the crankcase and the combustion chamber, and as a very rich mixture has already been prepared in the crankcase it rushes into the combustion chamber, owing to the suction already present there, as soon as the piston uncovers the passage ports on its downward path. The upward stroke of the piston compresses the mixture of pure air from outside the engine and very rich mixture from the crankcase in the ordinary manner, and, as in the previous example, the third stroke, i.e., a downward movement of the piston, provides the power as the mixture is fired at the top of the stroke. On the fourth stroke, the valve opens again and allows burnt and used gases to pass out into the air; then the cycle commences once more, the valve remaining open, as already explained,

to allow pure air to enter the engine during the first half of
the downward inlet stroke.

One Principle, Many Types

From this very elementary description of the principles
on which the petrol engine works the pupil may have some
difficulty in recognizing the various parts when confronted
with a modern aeroplane engine.   He will find, however,

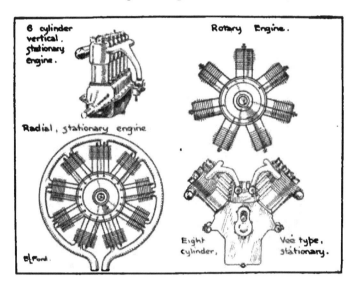

Four types of aeroplane engines.

that if he has a sound idea of the fundamental principles of
the working of a single-cylinder motorcycle engine, and if
he can take it to bits and put it together again he will soon
pick up the peculiarities of the more complicated aviation
engine.   Perhaps the arrangement of the cylinders on the
crankcase may strike him as unusual, or it may be that the
position of the valves or methods of driving the camshaft

puzzle him. But familiarity with the first principles will soon show how they work.

Aeroplane engines may be classified as follows:—

Stationary Engines

Cylinders arranged in line vertically as in motorcar. Cylinders arranged in V formation.	Example:—160 h.p. Beardmore, six-cylinder, vertical. Example:—80 h.p. Renault or 90 h.p. R.A.F., eight cylinders in two sets of four, set in the form of a V.
Cylinders arranged radially.	Example:—100 h.p. Anzani, cylinders set radially round the crankshaft.

Rotary Engines

The cylinders rotate round a fixed crankshaft.	Example:—80 h.p. Gnome; 80 h.p. Le Rhone; 100 h.p. Monosoupape; 110 h.p. Clerget.

The arrangement of the valves in these engines varies. The commonest practice is to fit the overhead type; on the 80 h.p. Renault and the 90 h.p. R.A.F. the exhaust valves are of the overhead type and the inlet valves are placed under them. The camshaft is carried between the cylinders at the bottom of the V and lifts the valves by tappets. On the Curtiss, both valves are overhead; on the Gnome, the inlet valve is, as already described, situated in the piston, and is of the automatic, or atmospheric, type. The exhaust valve is carried in the cylinder head, and is driven by cams and tappet rods. On the Monosoupape, the overhead valve serves both as exhaust and air intake valve. On the Le Rhone and Clerget engines, both inlet and exhaust valves are carried in the cylinder head, and are operated by means of tappets. Some different valve arrangements are illustrated.

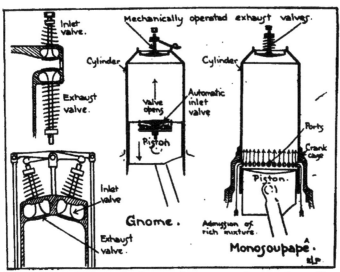

Arrangement of inlet and exhaust valves on aeroplane engines. Top left: The super-imposed arrangement. Bottom left: Overhead inlet and exhaust. Centre: Gnome pattern inlet in piston, and over-head exhaust. Right: Monosoupape or single-valve engine.

Generally speaking, the rotary engine is lighter per pound per horse-power than the stationary type. Rotary engines are all air cooled, whilst only the smaller stationary engines follow this practice. On the other hand, the rotary engine is often more extravagant in petrol and oil consumption than the stationary type, and this has to be considered in weighing up the relative advantages and failings of each type. Generally speaking, rotary engines are used on most small and light machines, such as scouts, which are single-seaters. On passenger-carrying machines, used for long distance work, a stationary engine, generally water cooled, is used on account of its reliability and economy.

It can now be assumed that the pupil has at any rate an elementary knowledge of the principles on which the engine works, and he desires to pick up as much practical knowledge about it as he can in the few weeks he has at his disposal.

Parts that Give Trouble

Whenever an engine fails or is being taken down for repair the pupil should find out what has happened to it and what is being done to remedy it. The parts of an engine most liable to give trouble (excluding serious mechanical breakdown) are the sparking plugs, the ignition wiring, valves and valve springs, the magneto, the carburetter, the petrol pipe, the inlet pipe, piston rings, obturator rings (where fitted), and the lubrication system. By taking these parts one by one and discussing the symptoms, causes and cures of the troubles, the pupil may learn the kind of breakdown to expect, and will see how, by a little care on his part, with an occasional inspection of the machine, he can prevent many of these failures.

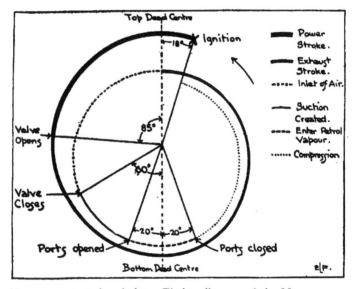

Monosoupape engine timing. Timing diagram of the Monosoupape engine, showing how the single valve acts as air inlet and exhaust. The rich mixture is admitted from the crankcase by means of transfer ports in the cylinder walls.

Engine Troubles, Symptoms, Causes and Cures.
The Plugs

Sparking plugs are somewhat apt to become sooted up with burnt carbon between the plug points, or with oil getting up into the body of the plug. In the latter case, the plug points may appear clean and yet the plug may be faulty. When in doubt, change the plug, and if the cylin-

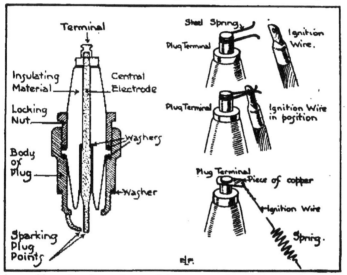

A sparking plug in section and two types of springs on plug terminals in common use on aeroplane engines.

der then commences to fire the plug has obviously been the cause of the trouble. A new washer should be fitted with each plug when it is changed. Sometimes a clean plug may be faulty also, so that unless the two plugs have been tried do not take it for granted that the trouble is elsewhere. Plugs are sometimes capable of being taken apart and cleaned with a stiff brush and petrol. The points should be cleaned and set the correct distance apart, which is approximately 1-50th of an inch. Sometimes, owing to quick changes of temperature, water will form on the plug points and cause misfiring. Sometimes the insulation of the plug will be found to have cracked owing to excessive heat, in which case cleaning is useless and the plug must be scrapped. Misfiring may occur owing to the plug terminal dropping off through being badly secured. Spring terminals, if reliable, are the best, as otherwise a great deal of time is wasted in changing plugs. Mechanics should carry spare plugs and a plug spanner in their pockets for the same reason.

Quick-release Terminals

Two types of spring terminal are illustrated (page 30). The first consists of a piece of bent wire fitted round the neck of the plug, with its two ends cranked and sprung slightly apart. The insulated wire from the magneto is fitted with a copper terminal with a hole in it through which the ends of the plug spring terminal are inserted by being pressed together. They then spring apart, and there is no chance of the insulated wire dropping off the plug, although it is only the work of a moment to pinch the terminal ends together and release them when desired. The second type of spring terminal incorporates the high-tension lead wire as well, and is used on engines such as the Gnome, Clerget and Monosoupape, where there is no need to employ a rubber insulated high-tension wire. A steel wire is used to convey the current from the distributer to the plugs, and incorporates a spring in its design. The sketch will explain the idea. To detach the wire it is only necessary to take hold of the loop at the end and strain the wire against the spring, thus relieving the pressure of the upper loop round the groove in the plug. If a plug constantly oils up at its points it may be found expedient to increase the gap between the points considerably. This, however, may make it difficult or impossible to start the engine on this cylinder, but it will be possible to start it on one of the others, and the cylinder in question will chip in when the engine is running. Moisture on the external insulation of a plug will also cause misfiring, which will cease if the plug is dried.

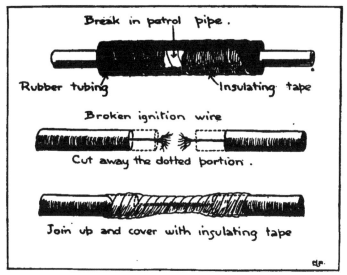

Break in petrol pipe.

Rubber tubing · Insulating tape

Broken ignition wire

Cut away the dotted portion.

Join up and cover with insulating tape

Repairs on the field. Top: How to repair a broken petrol pipe. Bottom: A repair to a broken or chafed high-tension wire effected by means of insulating tape.

High and Low Tension Wirings

The ignition wiring does not give so much trouble as do the plugs, but it should be examined from time to time to see that it is not chafing against any rough parts of metal or wood. The insulation material is rubber, and this may easily perish if it is constantly subjected to oil or friction with some part of the machine. Wiring should be neatly clipped up in position by fibre clips, and should not lie in long unsupported lengths, as this strains it and may lead to breakage owing to vibration. The wiring is divided into two systems, the low tension, which requires a thinner insulation than the other, which is the high tension. The low-tension wiring runs from the low-tension terminal on the contact-breaker cover on the magneto to the switch, whilst the high-tension wiring runs from the magneto to the

distributer, and from the high-tension terminals on the
distributer of the magneto to the sparking plugs. The high-
tension wiring is generally thicker than the low-tension.
To repair a chafe in ignition wiring, black insulating tape,
or ordinary medical tape, can be bound round the weak
part. If the wiring breaks the insulation can be cut off
so as to expose the wire core for a short distance. The two
ends can then be twisted together and the whole joint bound
with insulating tape. The reason that in some engines,
notably the Gnome, the high-tension wiring is left free is
because it is led direct from the magneto to the sparking
plugs, and so touches nothing en route that would cause a
short circuit. In this case the air acts as an insulator.

Switches

The switches require to be continually looked after to
see that they are working efficiently. The tumbler type,

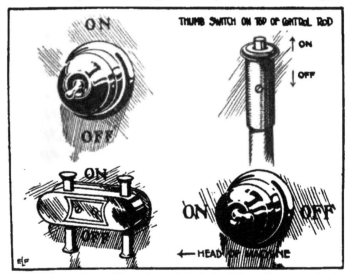

Types of switches. When the switch is "on" the pilot or mechanic
calls out "contact"; in the "off" position the phrase used is
"switch off."

such as is used in electric lighting installations, is the most reliable. The spring button type fitted on the top of some control levers is liable to stick at times, and should not be depended upon, as the spring is often uncertain in action. In this type the switch has to be held down for "switch off," and springs on to "contact" when released. All switches are supposed to work in the same direction and to be marked for "contact" or "switch off," but this is not always the case, and the pupil should make certain that he clearly understands how the switch works. In the tumbler type the switch springs on to contact easily but requires rather more force to put it into the "off" position.

Valve Troubles

Valves are subject to many different forms of trouble. Generally, it is the exhaust valve which fails and requires more attention, because it is subjected to much greater heat and strain than the inlet. When a valve breaks off at the head the pilot should switch off and land, as otherwise there is every chance of the valve head entering the combustion chamber and breaking up the engine. The symptoms in a broken valve are the sudden misfiring of one cylinder, probably accompanied by a sharp metallic knocking noise in the exhaust. The cure is to remove the broken parts and to fit a new valve. Valves should be changed periodically, and an examination made systematically will often insure the removal of a valve that would break after a few more hours running. If the valve is burnt and cracked under the head it should be changed at once. Valves also warp, owing to the heat, which means that they do not bed down evenly on their seating. This makes it impossible to secure a gastight joint. The cure is to re-cut or re-grind the valve. Valve grinding is an art in itself, and the pupil should familiarize himself with the methods employed. The object of valve grinding is to make the valve rest evenly on its seat. When the valve has been properly ground-in and cleaned up there should be no black spots or rings on it or on its seating. Valve grinding is done by means of a screwdriver engaged with a groove in the head of the valve and a valve-grinding compound. A spring is placed under the valve head so as to raise the valve from the seating from time to time, and it is bad practice to grind the valve without taking this precaution. As little of the grinding compound as possible should be used, as otherwise the seating will grind into the rings. Valves may also stretch, which fault can be watched for by noting the clearance between the stem and the rocker arm or tappet rod. Stretching is often a forerunner of the complete breakage of the valve, and is due to great heat or to the valve spring being too strong. Any squeaking that manifests itself in the valve guide or rocker arms shows that they are dry and that friction is taking place. The cure is generally to lubricate them or treat with graphite. If the valve stem is bent or distorted it will also bind in the guide. The cure is to fit a new valve, as straightening the old one is difficult and unsatisfactory. It is often difficult to fit new valve springs, which have to be compressed before they can be got into place, but if they are first of all compressed in a vice and tied up in position they can then be fitted quite easily.

Ignition Troubles

The magneto is perhaps too complicated a piece of mechanism for the pupil to take to pieces and repair on the field, but a few simple breakdowns which can easily be remedied may be described. Some engines employ two magnetos, each firing half of the number of cylinders. If one magneto fails, the pilot can soon find out which it is by feeling which cylinders are cold and then tracing the high-tension wiring from the cold cylinders to the faulty magneto. To test for a faulty magneto, leave the switch on, put one cylinder in the firing position and rock the propeller gently to and fro. Feel the high-tension terminal of the plug, and if no shock is experienced it proves that the magneto is at fault. The greatest enemy to the efficient working of a magneto is oil and dirt. Therefore, very often, merely cleaning up the magneto will cure the trouble. The distributer should be free from oil (which it

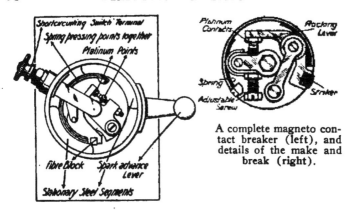

A complete magneto contact breaker (left), and details of the make and break (right).

very seldom is on rotary engines), the contact breaker should be clean, and the high-tension terminals and brushes should be cleansed carefully to get rid of any lubricant which has crept in. The brushes should work freely in their guides or holders, and their springs should be strong enough to ensure efficient working. Frequently, trouble is encountered in connection with the contact breaker. The platinum points may require cleaning, trimming and adjusting to their correct clearance or else, especially in damp weather, it may be found that the rocker arm, which operates against a cam and opens the points, has seized. The cure is to ease it in its bush and vaseline the metal pivot. This trouble may either cause total or partial misfiring, the latter being by far the more difficult to detect. Broken brushes cause total misfiring, while dirty distributers or contact breakers are frequently indicated by partial or intermittent misfiring or difficulty in starting. Sometimes, in damp weather, moisture condenses inside the magneto and causes partial and erratic firing. To cure this trouble, the magneto should be taken to pieces and the armature and magnets dried in a warm room. See that a bar of iron connects the two ends of the magneto, as they may lose their magnetism. Moisture on the high-tension terminals will also cause misfiring, which will disappear if the terminal is dried.

Failure in Petrol Supply

Carburetter troubles are mostly experienced in connection with the jet, which every pupil should be able to remove. A jet key should be part of every pilot's tool kit, as it is often impossible to remove the jet with the ordinary type of spanner. Popping and uneven firing indicate partial restriction in the supply of petrol. This may take place at the jet, where the flow of petrol from tank to engine is naturally most restricted, in any of the small petrol passages of the carburetter, or in the petrol pipe, filter, or petrol tap itself. All kinds of dirt seem to find their way

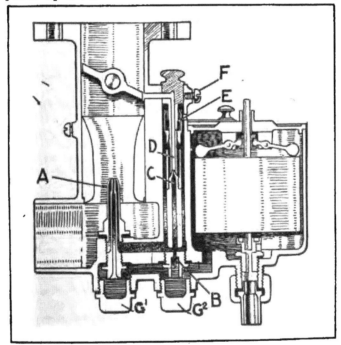

To illustrate carburetter design. (A) Main jet. (B) Compensating jet. (C and D) Slow-running jet and choke tube. (E) Air inlet. (F) Set screw holding compensator. (G^1 and G^2) Screw caps below the two jets.

into the carburetter from time to time, and it is a mystery where they all come from. Cotton wool, grit, bits of rubber from petrol joints or composition piping would all cause stoppage. If a failure of the petrol supply is suspected and the tanks are full, the first thing to do is to turn on the petrol and flood the carburetter. The petrol should then pour out; if it does not, the petrol union should be disconnected at the carburetter and the flow tested again. If it is good the fault is in the carburetter. If bad, the trouble lies in the pipe, filter or tap. Should the carburetter be assumed to be at fault, remove the jet, see that it is clean and blow down it. Clean out the float chamber with a cloth and petrol, re-assemble the parts, and test the carburetter again. In cases where the pipe supply is restricted, take the pipe off from the tank and blow down it until it is free. If the pipe is not blocked, turn on the petrol tap at the tank and watch the flow. If it is poor, clean out the tap and tank, for the root of the trouble lies there. In the same way the filter should be tested and the gauze thoroughly cleaned. Specially made petrol pipes of composition often perish after a time, and should therefore be renewed before they disintegrate and cause choking. Petrol pipes should never be unsupported for more than short lengths as the vibration of the engine will cause fracture. A rubber joint is often inserted in a long length of piping to overcome this risk.

Repairing a Broken Petrol Pipe

To repair a broken pipe rubber tubing can be slipped over the adjacent broken ends, which are then bound up with copper wire and insulating tape. A punctured float will also cause trouble in the carburetter. This can be detected by removing the float and shaking it. If there is petrol inside, fit a new float and return the other to the mechanics, who will heat it, thus vaporizing the petrol, which will emerge in bubble form if the float is immersed in water, and thus indicate the position of the puncture. The float can then be soldered up and made air-tight again, if care is taken that the addition of the solder does not weight it and so upset the level of the petrol in the jet.

The correct mixture is indicated by a perfectly-running engine, and if misfiring occurs when the ignition is correct it will indicate too rich or too weak a mixture. Black smoke from the exhaust indicates too much petrol, as does a long yellow flame. Too rich a mixture may be due to flooding of the carburetter at the jet, or shows that the air valve is stuck and remains closed. A restricted supply of petrol or air leaks in the carburetter or induction pipes, will result in too weak a mixture being formed. In winter, larger jets are often used than would be necessary in the summer. In cold weather it is possible to facilitate starting by heating the carburetter and inlet pipes with a hot rag to assist vaporization.

Difficulty in Starting

The main air ports to the carburetter may also have to be closed during induction in order to prepare a very rich mixture for starting up. If the engine and carburetter are water jacketed, hot water can be poured into the radiator, and this will materially assist in starting the engine. Sometimes, although the petrol may have been filtered through gauze or chamois leather, water will be found in the jet and float chamber. Unless it is removed misfiring will occur. Petrol or oil funnels and filler caps should never be left on the ground, as they pick up dirt, which will probably enter the petrol and oil tanks and cause trouble. On an engine such as the Gnome, where no float chamber is fitted, these remarks apropos float chamber troubles obviously do not apply. In this latter type of engine the only thing that can go wrong is a petrol jet situated in an inlet pipe through which air is taken into the crankcase. The regulation of the petrol supply, which is by hand instead of by float, is, however, a much more difficult matter. Sometimes the air control lever or throttle wire may break and cause the engine to stop. If the return springs of these levers are set to keep the throttle and air valves open in case of failure of the engine controls, this trouble will be overcome.

Inlet pipes can only cause trouble by cracking at their joints or breaking away at the unions. In either case the

with the tail down. Sometimes, in order to warm up this type of engine more effectively, the tail is jacked up to a height of several feet while the machine is in the flying position. The oil will then be distributed evenly throughout the crankcase and will reach the pistons and bearings more quickly. In any case it is advisable to warm up the engine gradually and slowly from cold, and never to run it at maximum revolutions for longer than is absolutely necessary. This applies not only when testing the engine on the ground, but also when flying. Very seldom, except in climbing, is full engine power desired or needed. Level flying can be carried out on most machines with the engine running at several hundred revolutions below its maximum.

Common Engine Failures

Should the engine stop suddenly the cause will be failure of ignition or of the fuel supply. To cure it, test the magneto and switch, and then the petrol supply.

The misfiring of one cylinder will probably be due to a faulty plug. If the misfiring is accompanied by a loud banging, or rattling, the cause is probably a broken valve.

If one set of cylinders misfires the reason may be put down to magneto failure.

If irregular or infrequent misfiring occurs it will be because the rocker arm on the magneto contact breaker sticks occasionally, or because there is oil or dirt on the distributer, or the platinum points require trimming.

Misfiring, accompanied by spluttering, is brought about by an insufficient supply of petrol or the air valve sticking.

Difficulty in starting the engine may be due to the cold weather, or the fact that the plug points are set too far apart. The magneto contact breaker or distributer may require cleaning, or there may be an air leak in the induction pipe. If the engine runs well when it has once been started the last cause may be suspected, as the mixture might be too weak for starting but correct for general speed.

Black smoke from the exhaust may be the result of too rich a mixture, or show that the carburetter is flooded, or an air valve stuck in the closed position.

Run a New Engine Gently

Should the number of engine revolutions drop after it has been run for some time, without misfiring, partial seizing up of bearings or pistons has probably occurred, as the result of either faulty lubrication or lack of play in the parts concerned. This is likely to occur on a new or recently overhauled engine. Great care should therefore be taken during the first few hours flight of a new engine in order to allow the bearings and other parts to run in. If the steel cylinders are blued, or discoloured, the obturator ring may be suspected, if one is fitted. In passing, it may be mentioned that an obturator ring is a special form of piston ring used on rotary engines in conjunction with the ordinary type of ring. It is made of brass, and is L shaped, with the ordinary piston ring between the piston and the downward stroke of the L. It is necessary owing to the uneven cooling of the rotary engine, the front of the cylinders receiving all the cooling draughts.

If the engine runs well for a few seconds, then splutters and continues to run, the cause may be a partially choked jet or water in the petrol.

Bad vibration of the engine is generally due to the misfiring of one cylinder, but it may also be caused by a loose propeller, or the engine being loose in the engine bearers. Sometimes the propeller is faulty, and this causes the engine to vibrate, in which case it should be changed at once. A small chip out of a propeller will upset its balance and will affect the smooth running of the engine.

General Hints on Running an Engine

Always warm it up slowly and gradually.

Never run it at full power, in the air or on the ground, for longer than is absolutely necessary.

Use as little petrol and throttle opening as is possible in accordance with flying conditions.

Remember that in case of engine failure it is always safer to make a landing with some cylinders firing than to continue until the whole engine gives up.

Listen to the sound of the engine and try and detect faults before they become serious.

Never open the throttle suddenly, as it strains the engine.

Run the engine on the chocks as little as possible, and on rotary engines do not use the "blip" switch more often than is necessary, as the life of an engine is greatly curtailed by non-observance of these rules.

Have the engine thoroughly overhauled at the regular intervals arranged for it.

On long cross-country flights ease the engine from time to time by throttling down or gliding.

Never go up without testing the engine, and never go up with a misfiring or unsatisfactory engine. Have any trouble put right on the ground.

CHAPTER IV

The First Lesson in the Air

PRIOR to receiving the first practical instructions in flying the pupil should realize that he cannot learn too much about the theory of flying and the means of controlling an aeroplane before he actually finds himself in the pilot's seat ready for his first lesson. He will learn very much more quickly if he has had already a preliminary grounding in the rudiments of mechanical flight than if he starts without a notion as to how or why a machine flies. Both instructor and pupil, therefore, will save themselves a good deal of time and trouble if the former explains to the latter the elementary principles of flight before attempting to instruct him upon handling a machine in the air. Emphasis should be laid on the particular importance of the method of controlling the machine and the mistakes to be guarded against in the air.

Dual-control Instruction

The method now almost universal of teaching pupils to fly is by means of dual-control machines. In the early days, after having acquired a theoretical knowledge of an aeroplane and the method of controlling it, the pupil was sent off for rolling practice, or "taxying," on an under-powered machine which could not possibly leave the ground. The next stage was to give him a machine with a wing surface sufficiently large just to allow him to get off the ground for short hops. From this he would advance to a machine which would just fly, upon which he would make "straights" until he was proficient, later on attempting turns on a more powerful machine.

This method has been superseded at all Service schools in favour of dual-control instruction. This means that the

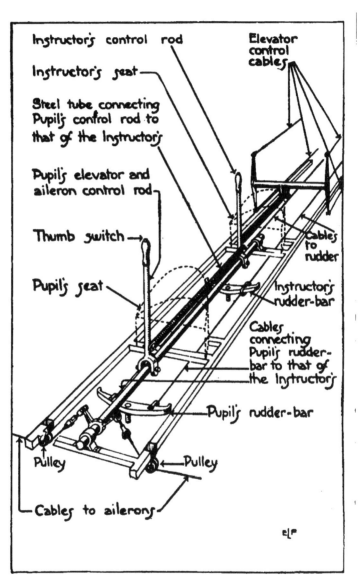

Instructor's control rod

Instructor's seat

Steel tube connecting
Pupil's control rod to
that of the Instructor's

Pupil's elevator and
aileron control rod

Thumb switch →

Pupil's seat

Elevator
control
cables

Cables
to
rudder

Instructor's
rudder-bar

Cables
connecting
Pupil's rudder-
bar to that of
the Instructor's

Pupil's rudder-bar

Pulley

← Pulley

Cables to ailerons

ELF

The duplicate controls of a dual-control machine. On different types
of machines the seating accommodation is arranged in different
ways. Sometimes the pilot sits in front and the pupil behind, and
vice versa.

machine is provided with two sets of rudder and control levers, one to be worked by the pupil, who generally sits behind, and the other by the instructor. On some machines one set of control levers is used, but they are placed in such a way that the instructor can work them by reaching over the pupil's shoulders. It is probably best for the pupil to be put in the pilot's seat from the beginning and for a telephone or speaking tube to be arranged between him and the instructor.

Communication in the Air

It is a great advantage if some means of communication be installed between the pupil and his instructor. A system of speaking tubes may be fitted or signs may be arranged. By this method a great deal of verbal instruction can be given in the air and the effect of the control movements demonstrated. This method obviously saves a great deal of time. If the instructor's position permits—as, for instance, if he were reaching over the pupil's shoulders in order to handle the controls—he may be able to speak to the pupil in the air without the need of a speaking tube.

Opinions are divided as to what is the best method and machine for teaching pupils. Some instructors prefer a "pusher" type, whilst others think that time is saved by taking a pupil on a "tractor" type from the beginning. In any case, the school machine should not be too fast or too sensitive on the control.

Working the Controls

There are many different methods of instructing, and it will generally be found advisable for the first lesson in the air for the instructor to take the pupil up to a calm and safe height, say 1000 ft. or 2000 ft., and then allow him to feel the controls. It is here that the importance will be appreciated of the pupil knowing what he ought to do before he has actually the chance of doing it. He should not be frightened of working the controls and no-

The tractor type of aeroplane with the propeller and engine in front of the machine. Note the long, tapering body or fuselage terminating in the tail. The tractor type of aeroplane is now employed practically universally, the old type pusher machine having been superseded even in school work.

48

The pusher type of aeroplane with the propeller behind the pilot.
Note the pilot's seat in the nacelle in front of the planes and the
tail booms and open frame work in place of the fuselage on the
tractor type.

ticing the effect they have on the machine. The instructor
may allow him to get into an unusual position and then
show him how easy it is to get out of it. It is far better
for the pupil to acquire the feel of the machine as quickly
as possible than to sit still and watch the instructor do
the work.

Flying Straight

The first lesson is to learn how to fly the machine straight.
Not only must the pupil fly in a straight line towards some
point which he selects in front of him, but he must fly the
machine level, i. e., neither upwards nor downwards. He
must fly laterally level, and not with one wing up and the
other down. He will find that, to begin with, he is unable

Flying straight. How to fly straight by selecting some prominent
object on the horizon, such as a church, and steering the machine
to it with the rudder. Another good method of flying straight is
to steer the machine along a straight stretch of road or railway, or
to fly it parallel with these landmarks. The machine is kept straight
by means of the rudder.

to control the machine in these three directions at once. He can get the directional straightness with the aid of the rudder bar, but he may allow the machine to roll sideways, or see-saw. On some machines it is necessary to keep on a certain amount of rudder in order to make them fly straight, and the pupil should realise that if he finds the machine drifting away to one side from the point he has chosen he should rudder in the opposite direction, and, if necessary, keep the rudder on.

Flying Level

In learning to fly level, i. e., neither climbing nor descending, the horizon should be watched and also the trim of the machine generally when it is set by the instructor in the correct position. The pupil can then attempt to keep the machine in that position by moving his control lever forward to prevent the machine climbing, or backward to prevent it descending. If he krfows the correct flying level speed of the machine, he can also watch his air speed indicator. He must remember that any increase in its speed (with the engine running at its normal number of revolutions) means that the machine is descending, and that any decrease in speed indicates that the machine is climbing. It is also necessary to bear in mind the fact that the air speed indicator is sluggish in action and only registers a change in speed after the machine has made that change. It should be borne in mind also that, if he is high, he will never come to any harm by allowing the machine to go too fast; whereas, on the other hand, he can easily allow the machine to get into an awkward position through permitting the speed to drop below flying speed, for it is the relative speed between the planes and the air that preserves the machine in equilibrium.

Horizontal Level

To keep the machine horizontally level it is necessary to watch the horizon and to attempt to maintain the front elevator (if there is one—or if the machine is a tractor, the wings) parallel with the horizon. This will only be

Trimming the machine to fly level (tractor type of machine) by
keeping the underside of the top plane parallel with the horizon. If
the pupil notices that one wing is down, he must move the control
lever sideways in the opposite direction to pull it up, and then
centre the lever.

In flying level on the Maurice Farman school machine, the front elevator is trimmed parallel with the horizon. This type of machine, which at one time was largely used for school work, has now been superseded by the tractor type. It is commonly known as the Long Horn, as opposed to the Short Horn, Maurice, which had no front elevator.

attained by practice, and the pupil should not be disheartened if he does not pick it up at once. When he notices that one wing is down, he should move his control lever in the opposite direction to that wing until the machine is flying horizontally again, whereupon he can move his lever back to the neutral position. In working any of the controls, it must be remembered to operate them gently, gradually and firmly. On no account should they be jerked or moved quickly.

Bumps in the Air

Every now and then what are termed "bumps" in the air are felt. These may affect the fore-and-aft or lateral level of the machine, and they must be corrected as they occur. The amount of correction necessary will only be found by practice. If a bump puts the front of the machine up, the pupil must move the control lever forward, and when flying level again he can centre his control. If, on the other hand, a bump sends one wing up, the lever must be moved sideways toward the wing that is up, in order to allow the other wing to rise; then, when he is level, he again centres his control. On some machines lateral bumps can be corrected by working the rudder in conjunction with the control lever. To do this he must rudder into the bump and work his control lever in the same direction. Thus, if a bump sends the right-hand wing up or the left-hand wing down, whichever way he likes to take it, the pupil must warp to the right with his control lever and rudder to the right also. When he is level again he moves his control and rudder into the central position as before.

Turning

The next stage, probably, will be to make a turn. This is quite simple. It is wise to begin a turn with a slight increase of speed, if the machine has a small range of speed. Both the rudder and warp, or the control lever, are then moved gently and simultaneously in the direction in which it is desired to turn, the right foot being pressed on the rud-

der to turn right, and the control lever moved to the right
to pull up the left wing. The machine will bank gently
and come round quite easily. Once the correct amount of
bank is on, the lever can be centred. When the turn is
completed the rudder and bank are taken off again, by moving the lever in the opposite direction, following up with opposite rudder and then moving the control and rudder to
their neutral positions. If the rudder is moved in the opposite direction too soon, the nose will go up and the machine
will slip inwards. The controls must not be centred again
until the machine is laterally level. This method applies
in general to gentle turns.

Before going on to describe how to do steep turns, the
pupil must grasp the action that the elevator and rudder
have on the machine. When the machine is flying level
and straight, a backward movement of the control lever
will always cause the nose of the machine to attempt to
reach the tail, and in doing this the nose will go up above
the horizon. The application of right or left rudder will,
in the same way, always tend to make the nose of the
machine try to turn to the right or left wing tip, so that,
when the machine is flying level, the rudder causes the
machine to turn, while the backward or forward movement
of the control lever causes the machine to climb or descend, or fly level, as compared with the horizon.

Inversion of Controls

Now consider what happens when the machine is made to
execute a vertical banked turn and is on its side relative
to the horizon. Take the case of a left turn. If left rudder
is held on when the machine is on its side in the turn, it
will cause the nose to approach the left wing, as in the
previous case, but, judging the position of the machine
from the point of view of the horizon, it will cause the
nose to drop below the horizon. Right rudder will have
the opposite effect, i. e., it will make the nose of the machine try to approach the right wing, which is uppermost:
in other words, comparing the nose again with the horizon,
right rudder will make the nose go up.

Next, take the action of the elevator when the machine

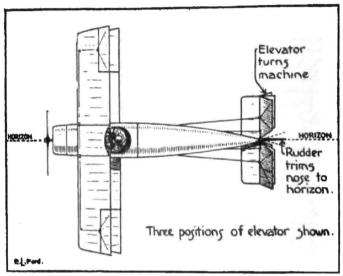

Illustrating the inversion of the controls on a vertical turn.

is on its side in a steep turn. If the control lever be pulled back, it will cause the nose of the machine to try to approach the tail, but as the machine is on its side, in relationship to the horizon, this backward movement of the stick will simply speed up the turn and make the machine turn more quickly. So that if the nose of the machine is still compared with the horizon, it is found that on a vertical bank the nose is trimmed above or below the horizon by the use of the rudder and the machine made to turn by the use of the control lever which operates the elevator. At turns of 45 degrees, the rudder acts half as rudder and half as elevator, and the elevator acts half as rudder and half as elevator. At lesser angles the elevator is more elevator than rudder, and the rudder more rudder than elevator, when the nose of the machine is compared with the horizon, whilst at steeper angles of turn the elevator becomes more and more the rudder, and the rudder more and more the elevator.

Now to describe the method of making a steep turn, the control movements for which the pupil, having mastered the idea outlined in the preceding paragraph, will not find at all difficult to understand. To make a steep turn to the right (or left), bank and rudder are applied in the desired directions; in this case to the right, until the machine is more or less on its side. Then the nose of the machine is trimmed to the horizon by means of the rudder, right rudder bringing the nose down and left rudder bringing it up. It will be found necessary to ease off the right rudder while the stick is then pulled back so as to make the machine turn. While the stick is pulled back, it will be found

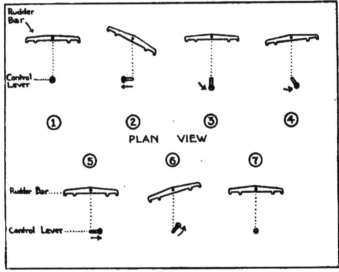

Movements of the rudder and control lever on a steep left-hand turn (plan view). (1) Control lever and rudder in neutral or central position. (2) Left rudder and left control lever—machine takes up bank. (3) Rudder eased off to trim the nose of the machine to the horizon, and control lever brought back to make the machine turn. (4) Rudder eased off still more if necessary, and the control lever brought to opposite side and fully back to prevent overbanking. (5) Coming out of the turn—control lever moved over to the right first. (6) Then, as the machine approaches the horizon, the control lever can be moved forward and right rudder applied. (7) When the machine is approaching the horizontal both levers are moved centrally.

that the bank of the machine is inclined to get steeper and
steeper owing to the increased speed and consequent lift
imparted to the left wing by its controlling aileron which
is down, so that it is then necessary to ease off the bank
slightly; that is, the stick, besides being pulled back to speed
up and make the machine turn, must be moved towards the
opposite side, viz., to the left. Once the correct position
of the levers has been found the machine will continue
to turn round and round without losing height or increasing
its bank or radius of turn. To get out of the turn, opposite bank is used; that is, the stick is moved to the left,
and, when the machine is approaching its horizontal and
lateral level again, left rudder must be applied and the
stick can then be moved forward again to its central position. Gliding and climbing turns are made in exactly the
same manner, except that, in the former case, the nose of
the machine must be trimmed below the horizon, and in the
latter above the horizon, the amount below or above the
horizon depending on the speed and rate of descent or
ascent required.

Learning to Land

In passing, it may be noted that there is no need for the
pupil to hold the controls as tightly as he can, or to jam
the rudder with his feet. A machine can be controlled with
one hand holding the control lever quite lightly; in the
same way, the rudder can be operated by a mere touch of
the toe.

When the pupil can fly the machine correctly in the air,
he will be taught to land and get off. Many instructors
give landing practice by flying round and round in circles
and landing once in each circuit. This method is all right
if the machine is brought to a standstill after each landing,
and a few words of advice are given to the pupil as to the
mistakes he made in his previous get off and landing. An
instructor can teach a pupil as much, if not more, by talk-
ing to him and taking him into his confidence than by going
on flying for hours without verbal instructions. The per-
sonality of a pupil should be studied if the best results are
to be obtained from him. In giving landing and getting-off

instructions, the faults made on the previous circuit should be explained. If a pupil does not understand why the machine behaved in such and such manner when he last landed or got off, he should ask the instructor the reason. The instructor would also be well advised to ask the pupil at the end of each flight if he has any questions. Even if he has not, a few words may be given in explanation of any point that has arisen whilst in the air. All this shows the advantage of some means of communication being arranged between pupil and pilot if the flying lessons are to be concluded with the greatest expediency.

It may be pointed out that, to begin with, short flights of 10 to 15 minutes duration are of more value to the pupil, who easily becomes tired until he gains experience, than longer flights of from 30 to 60 minutes.

Getting Off and Landing

In getting off, a pupil must accelerate the machine gradually, keeping it straight towards some point by means of the rudder, and when it has attained its flying speed pull back the control lever or elevator very gently. Generally, when the machine is just in the air, he should ease the lever forward momentarily to allow the machine to gather more speed. He can then continue his climb with a margin of safety in hand for emergencies.

In landing, the pupil first cuts off the engine, puts the machine at the correct gliding angle by moving the control lever forward, then, on approaching the ground, which he must watch intently 20 or 30 yards ahead of him, he pulls the lever back slowly until he has flattened out a foot or two above the ground. If he holds the lever still, the machine will slowly sink as it loses its flying speed and touch the ground quite slowly and without any jar. In landing, it may be necessary, owing to bumps or misjudgment, to ease the control lever forward or to hold it for a moment, but the general tendency will be a backward one.

Some further notes on getting off and landing will be found in the chapter which deals with the pupil's first solo flight.

When to Make a Solo Flight

The moment when a pupil is ready to go solo is a point of importance both to him and to the instructor. The latter has the responsibility of a possible accident if the pupil is sent up too soon; while the pupil has his own neck to consider and the damage he may do to an expensive machine by making a bad landing, or committing some other error, which ends in an accident, through being over-confident or too keen to go solo. The time a pupil takes to learn depends on the type of machine upon which he is taught, his capabilities, and also the instructor. Some men learn to fly in 2 hours; some take as long as 10 or 12 hours.

An instructor should also ask the pupil if he feels confident and would like to go up solo. If the reply is in the negative, then, generally, more instruction must be given. The pupil must have confidence in his ability to handle the machine before he goes up alone, which will be increased if he has already gained a certain amount of faith in his instructor as a judge of his own capabilities.

Before a pupil is allowed to go up alone he should be able to land, take a machine off the ground and fly it in the air for six short circuits without the instructor touching the controls.

If the pupil feels confident, and the instructor is satisfied as to his capabilities, he should go up solo, preferably on the identical machine on which he has been instructed. To put the pupil in another machine, where, perhaps, the instruments are placed differently, and where the adjustment of the machine may be different, even though it is of the same make, is to handicap him. Again, to teach a pupil in the front seat of a two-seater and then to send him up solo in the back seat, or vice versa, is to take a risk which is quite unnecessary, for the positions of the pupil and instructor can easily be changed on most dual-control machines. A pupil is thus given every chance of making a good first flight. He should be sent off, if possible, with little or no warning, in order that he may be caught with his blood up, as, for instance, immediately after an instructional flight. Some pupils are most imaginative, and if they have no time to ponder over the possibilities of their first solo flight, they take the air much more readily and confidently.

The first flight should be on a calm day, preferably when there are no other machinès in the air. Exact instructions as to where to fly and the length of the flight should have been received beforehand. Generally, four circuits and landings are enough for the first solo. The pupil should have been taught at what height, and over what landmark, he should. switch off in order to land at the desired spot in the aerodrome. Having satisfied himself that the machine is in proper condition, the instructor will do well to withdraw so as to let the pupil get into the air without any nervousness that may be caused by the thought that the instructor is watching him.

Instruction on Advanced Types

When a pupil can fly one type of machine really well there is little need for a great deal of further instruction on the more advanced types. For this reason it is better to keep á pupil on one type of machine for several hours until he can fly it really well than to send him up on other kinds of machines, none of which he has had time to fly properly. Given sufficient room on the aerodrome for fast landings, a man who can fly one type of machine well should be able to fly most others with tolerable success. He will only master special peculiarities of different types of machines by experience. Probably he will notice that the speed and sensitiveness of the controls are the chief points of difference.

A great advance has been made recently in dual-control machines, some fast machines having dual control nowadays. There is, therefore, no need to take unnecessary risks by sending pupils up solo straight away on these machines. It is better to work up from slow and sluggish machines to fast and sensitive ones, with a little preliminary dual control on each. It is only the landing and getting off that require attention, and half-an-hour's instruction on these points should be sufficient. The pupil should remember, however, that he is not entitled to as much instruction on the more-advanced types of machines as when he was learning to fly. In the early days, when there were only one or two types of dual-control machines available, as soon as pupils could fly them they had to go on to the more-advanced types without any instruction whatever.

CHAPTER V

The First Solo Flight and Aerodrome Practice

BEFORE starting on a flight, whether the first solo or a 200 miles cross-country journey, it is a wise plan for a pilot to look over his engine and machine methodically. He should not take it for granted that the machine is ready to fly because the mechanic reports it so, but must look over it himself. Some pilots have the mechanical instinct born in them; others have not, and will never attain it. The latter are the people who are in the habit of using pliers to undo nuts, or who employ wood chisels in the place of a screwdriver.

Attention to the Engine

First glance over the engine, which is the more important part; and here a thorough knowledge of aero engines will be of great value. The following are some of the details that should be examined:—The terminals on the sparking plugs and the magnetos should be tight; both the high and low-tension wiring should be neatly clipped up and not lying about the machine in festoons, which might tend eventually to chafe through and cause a short circuit (at any points where wiring touches the metal portions of the machine, some form of additional insulator, such as insulating rubber tape or a fibre ring, should be inserted); the petrol unions should be tight; rubber connections, if fitted, must not be perished; and if the petrol pipe is made of some woven material it should be examined for signs of disintegration. Tappets and rocker arms should be inspected, and both should be properly lubricated. It should be noticed if any split pins are missing, and that those that are fitted are not loose. If automatic inlet valves are used, the pupil should see that they are working properly.

63

Finally, the pilot should be satisfied that there are no petrol or oil leaks, and that the switch is working properly.

Overlooking the Machine

Next, the machine itself must have some attention. After some hours flying experience on one type of machine, the pilot will know which parts are most liable to give trouble. These may be the wires slacking off in the tail, or the tail itself may be liable to work loose. Wear should be looked for in the Sandow shock absorbers in the under-carriage, or perhaps on an aileron which is inclined to bind. Loose or missing nuts and bolts should be looked for. The turn-buckles should be correctly locked and the control wire pulleys or guides greased properly. They should work easily without too much slack in the wires. Of course, the controls are most important, the most vital being the elevator control. They must be tested by sitting in the machine, working the control lever and rudder bar, and, at the same time, observing if their operation has the desired effect on the ailerons and tail. If the wires show signs of chafing where they pass over the pulleys, they should be changed. In some machines it is possible for the mechanics after an overhaul to connect the elevator control wires in the wrong way, i. e., they may forget to cross them. It is thus necessary to pay particular attention to the machine after it has been in the shops or when the controls have been over-hauled. The nuts and bolts and the locking arrangements for securing the landing chassis wheels should be inspected occasionally.

Preliminary Engine Tests

Whilst sitting in the machine the pilot should test the engine and see if it is developing its proper number of revolutions; if not it requires attention. This is an important detail, because a slight falling-off in engine revolutions means a very great reduction of the power developed. At the same time, it should be remembered that an engine will often give 50 or 100 more revolutions per minute in the air than it does on the ground, and this must be allowed for. A skilled pilot can tell, of course, by the sound of his engine whether it is missing fire or running unevenly or weakly. He should then switch off and tell the mechanics to feel the cylinders, the cold one being that causing the trouble. A partial misfire is always more difficult to locate than a total misfire. The latter must be due to one of two things—either a choked or broken pipe, or faulty ignition at the magneto or switch.

Knowledge of the Engine

Too much importance cannot be attached to the need for the pilot to understand his engine and machine perfectly. When he knows his engine well, and has the feel of the machine, he can tell exactly how it is flying, and what is more, he can often discern faults in the machine or the engine before they become really serious. If his engine fails, he should be able, after some hours flying, to diagnose the trouble before he comes to earth, just in the same way as a good motorcar driver can tell what has caused his engine to stop before the road wheels of his car have come to rest.

In warming-up an engine it is wise, when cold, to open it up slowly, in order to allow the metal to expand evenly and to permit the oil to thin and circulate properly through the bearings before full power is applied. In cold weather a longer period must be given for warming-up than in summer time, when the oil is not congealed. In very cold weather the castor oil used on many machines will solidify almost to the condition of candle grease. To thin it, methylated spirit may be mixed with it in the proportion of one pint to a gallon of oil; petrol can also be mixed with the oil in the proportion of 1 in 20.

The pilot should remember that the oil supply must be turned on before starting the engine, otherwise it will seize up very soon. On a rotary engine he must note the rise and fall of the oil contained in the pulsator glass domes. If these pulsations show up, he can take it that the oil pump is working and the oil supply turned on.

Starting Up

The engine and machine having been tested, the pupil is now ready to start. It should be noted that the onus of accidents occurring when starting up the engine by swinging the propeller rests with the pilot. Incidentally, the mechanic should always treat the propeller as if the engine were "on contact," and, for this reason, it is a good plan for the switch to be fitted on the outside of the machine where the mechanic can see it.

When ready to suck the mixture into the cylinders, the mechanic shouts "Switch off!" or "Suck in!" whereupon the pilot, first looking to see that the switch is off (it should have the "On" and "Off" positions clearly marked), replies, "Switch off." The mechanic then sucks in until

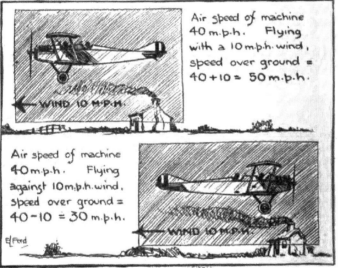

Air speed of machine 40 m.p.h. Flying with a 10 m.p.h. wind, speed over ground = 40 + 10 = 50 m.p.h.

Air speed of machine 40 m.p.h. Flying against 10 m.p.h. wind, speed over ground = 40 - 10 = 30 m.p.h.

WIND 10 M.P.H.

WIND 10 M.P.H.

The importance of wind. The air speed of the machine is relative to the air, and not to the ground. If the whole atmosphere is moving at 10 m.p.h., and the machine air speed through the air is 40 m.p.h., the speed of the machine over the ground will be 50 m.p.h. Against the wind the ground speed will be 30 m.p.h. Hence the importance of getting off and landing against the wind.

Starting up without chocks. Men holding back the machine and keeping the tail down while the engine is tested.

immediately ready to start the engine, when he shouts, "Contact!" The pilot switches on and shouts, "Contact!" The mechanic then bumps the propeller over compression and the engine should start. The words "contact" and "switch off" are chosen so that there shall be no doubt as to what is meant, as there might be if the words "switch off" and "switch on" were used instead. Both mechanic and pilot should see that they speak distinctly and loudly. The switch must always be left off except when it is desired to start the engine. The importance of this can be realised when it is mentioned that a man's arm or wrist can be broken easily by the premature firing of the engine.

Importance of Wind

When starting, the pupil must remember to face the wind, and the same remark applies to landing. In both cases this is done in order to reduce the speed at which the machine leaves or lands on the ground. For instance, if there be a 10-mile-an-hour wind blowing, and the fowest flying speed of the machine is 40 miles an hour, by facing the wind the pupil can leave the ground and land at 40 minus 10, which equals 30 miles an hour. If, on the other hand, he starts and lands with the wind, he must do so at 40 plus 10, which equals 50 m.p.h., so that, in this case, it is seen there is a difference in the ground speed of getting off and landing of 20 miles an hour—a very considerable item when it is remembered it may be necessary to land in a small field where a limited space is allowed for the "run," or "carry," after the landing.

Signals to Mechanics

The usual method of signifying to the mechanics that the pupil is ready to get off is to wave the hand above the head, whereupon they will withdraw the chocks from under the wheels. When the chocks are withdrawn, the pupil will do well to see that all the mechanics are standing clear. It happens sometimes that one of them, possibly holding down the tail, has not seen the signal, and he or his clothing may easily become entangled in the machine if the pupil went off unexpectedly. There is another point which the pupil must guard against, and that is the possibility of the chocks being withdrawn in mistake before he is ready, or, alternatively, in a mechanic starting up the engine without first seeing that the chocks are in position. If the pupil were not already warned against these possibilities, he might easily run over the mechanic, owing to the forward rush of the machine, through the throttle being too wide open, or, in the case of a rotary type of engine, which cannot be throttled down so well as a stationary one, through keeping the switch on too long.

Having satisfied himself on these points, the pupil should make certain that he has a clear run in front of him and that there are no other machines in the neighbourhood, either getting off or landing, which can possibly baulk him. This applies especially to machines which might be landing over his head. If he has to taxi out some dis-

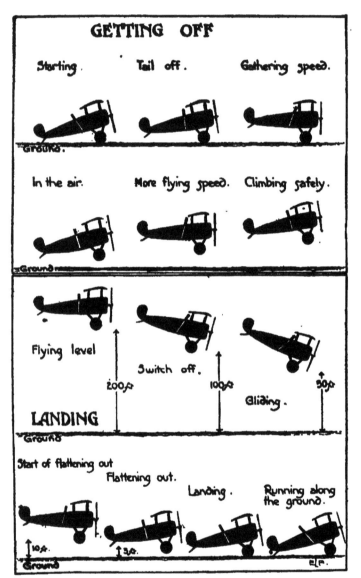

Getting off and landing. The movements of the control lever are indicated by the white line on the fuselage (exaggerated). It is of the greatest importance, both in getting off and landing, for the pupil to look ahead, so as to have some mark in view by which he can keep the machine straight. 69

Probable result.

Getting off with the tail too high, with the possible result that the machine might turn turtle.

tance in order to get into the wind, he should make certain that he allows himself ample room in order to clear such obstacles as woods, trees, or sheds.

Getting Off

Now comes one of the two more difficult parts of a pupil's initial flying experience—the start. It is a good plan to get under way slowly and to open up the engine gently, thus gathering speed gradually. Then, if the machine should tend to swing sideways it can be counteracted by ruddering in the opposite direction before the sideways swing develops into anything serious. A machine may swing sideways when starting for a variety of reasons. Sometimes it is due to the pupil, who is often liable to over-rudder on the ground until the machine is edging in the desired direction. This causes it to swing sideways fur-

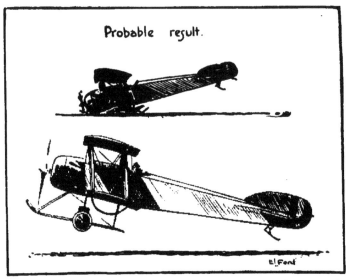

Probable result.

Getting off with the tail too low and without sufficient flying speed. Unless the control lever is immediately put forward, the machine will stall and crash.

ther than intended. The rudder should be eased off as soon as the machine starts to swing in the required direction. This applies more particularly to fast machines. Another possible reason for a machine swinging is that the pupil, in opening the throttle, which is placed on one side, may allow his foot to act automatically in conjunction with his hand, i. e., if he opens the throttle with the left hand he may push his left foot forward at the same time. When in doubt, it is a good plan to switch off altogether and to make a fresh start, unless the machine is already in the air.

Engine Economy

It is desirable to get under way gradually. This saves the machine, which should at all times be treated as a fragile and very sensitive piece of mechanism. It is well, also, to throttle down the engine a little in the air, for it is never a good plan to run it "all out" at any time.

On a long flight it pays to throttle down the engine every now and then and to glide down a thousand feet.

The pupil starts by moving his stick forward slightly as he gains speed, and by getting his tail well up, but not too high, for, if on a tractor, the propeller might hit the ground, gathering speed gradually and slowly and almost imperceptibly pulling the control lever towards him. When he feels the machine in the air he can ease the control lever forward momentarily in order to allow the aeroplane to pick up speed. The speed will then increase, but the rate of climb will decrease. Generally, it is wise for a pupil not to climb the machine at its lowest flying speed, but rather to allow three or four miles an hour for possible mistakes. The use of the instruments which indicate the speed of the machine is described in Chapter VI.

Turning

It is a bad plan at any time to turn near the ground because there is little air space in which to correct bumps, sideslips, or any other unexpected occurrences. It is wise, therefore, for the novice to keep straight on his course against the wind until he has attained a height of 500 to 1000 ft.; then he can make his turn.

When turning on school machines, where there is generally not a very wide range of flying speed or reserve of engine power, the nose of the machine should generally be put down slightly and the turn should be made at or slightly over flying level speed. Some instructors consider that the stronger the wind the more must the speed be kept up in turning. The reason for this, they say, is that in all turns a machine is inclined to lose height, and especially is this the case when turning down wind. Many pilots imagine that in turning down wind they experience a curious sloppy feeling in the controls, whereas they seem to detect a much firmer grip of the air when turning up wind. Much discussion and argument have taken place on this subject, but theoretically, at any rate, it should make no difference to the control of a machine whether it be turning up wind or down wind. How many pilots, for instance, could detect an up wind or down wind turn, if they could not see the ground, as would be the case if they were flying above the clouds? It is wise not to rely too implicitly at any time on the air speed indicator. This is specially true when turning, the danger lying in the possibility of the pupil holding off the bank when turning down wind near

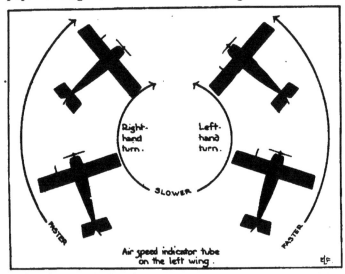

If the air-speed indicator is fitted to the left wing, on a right-hand turn it will register more than the mean speed of the machine. On a left-hand turn it will register less than the mean speed of the machine.

the ground, more particularly with stable machines. Assuming that the air speed indicator is fitted in the vicinity of one, say, the left-hand, wing tip, as is the case on most tractors, then, in making a turn to the right, the left-hand wing is obviously travelling much faster than the right-hand one. Therefore the air speed indicator tube attached to the left wing is over-registering the mean speed of the machine. On the other hand, if a left-hand turn is to be made, the indicator is under-registering the mean speed. On

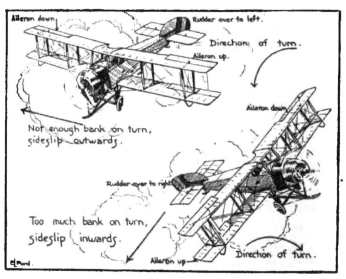

Incorrect turns, showing how sideslips occur.

tractor machines the air speed indicator is fixed to one side of the fore-and-aft centre line of the machine in order that it may not be interfered with by the slip stream of the propeller, which is driven out past the fuselage of the machine. In pusher machines, the air speed indicator tube is fitted centrally.

Straightening Up

In turning, a pupil should put his nose down slightly by moving the control lever forward. He should then push the rudder and control lever over together in the direction he desires to turn. When he wishes to fly straight again, he brings the rudder and the control lever back to the central position, or, in some machines, slightly over in the opposite direction, the opposite bank being applied first and the opposite rudder applied as the machine approaches a horizontal position: the quicker the turn the more rudder and bank are required. The controls are centred when the machine is level. He should never turn slowly, as this manoeuvre may result in a sideslip and stall, which means that the machine loses its flying speed and gets into a tail-down or cabré position.

The First Flight Solo

The pupil has now attained the height of 1000 ft. or so and is flying down wind. He is comparatively happy, as, once in the air at a fair height, the control of the machine is a simple matter. There are, however, several pitfalls into which he may tumble if he is not careful. He must always have in the back of his mind the possibility of an engine failure necessitating what is known as a forced landing. With this in view, he should always, so far as is possible, be able to turn into the wind and land in the aerodrome. Consequently, pupils who fly low over woods, houses, or other obstacles are taking unnecessary and foolish risks. If any useful purpose were to be served

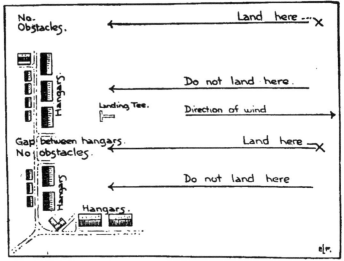

The landing ground of an aerodrome, showing where and where not to land. The same thing applies equally in getting off.

by running such risks, then by all means let him take them; but otherwise such flying is unjustifiable and extremely foolish. A pupil should remember that it is only given to the average pilot to make two or three such mistakes in the air before meeting with a really serious crash, attended by broken limbs, or perhaps worse.

For the same reason it is a bad plan to point the machine towards woods, trees, hedges or buildings unless the pupil is absolutely certain that there is sufficient space to prevent him from running into them in the case of an engine failure when starting, or to attain sufficient height first to enable him to turn round and land on suitable ground. In landing, too, he should allow himself as much room as possible for the run of the machine and should not head it towards obstacles, as he is likely to overshoot the mark considerably.

Flying Over Bad Ground

In certain circumstances, as, for instance, in the case of a cross-country flight in misty weather, when it is impos-

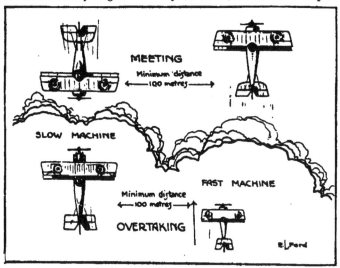

The rule of the air when overtaking or meeting another aeroplane.

Points to notice when encountering another machine in the air, i.e.,
avoid flying or landing in the backwash of another machine, as the
air in the wake of the machine is very bumpy.

sible to fly high, it may be necessary to fly low in the
vicinity of woods or other unfavourable landing ground.
Where possible, these grounds should be flown round, if
they are of great extent, instead of over them, as an engine
failure, with the ensuing forced landing in a forest, is not
an undertaking that a pilot relishes, although such landings
have been made in exceptional circumstances without the
pilot suffering so much as a scratch.

The Rule of the Air

Whilst in the air a sharp look-out should be kept for other
machines. This applies to all kinds of flying, but is particularly true at flying schools, where there is generally a
large number of machines in a comparatively restricted air
space. Usually, some set course is laid down, i. e., right
or left-hand circuits, and pupils should familiarize them-

selves with the course to be adopted before going up. In the air, the "rule of the road," as laid down by the Royal Aero Club, is that one keeps to the right when meeting machines and steers clear when overtaking. When machines meet at an angle the pilot that finds the other machine on his right must steer clear. In flying near other machines, a pupil may encounter bumps set up by the propeller of the other machine, but these need not alarm him, as it is a simple matter to correct or avoid them. For the same reason it is not advisable for a pupil to land behind another machine, as its wash may upset him considerably. In flying near airships he should beware of trailing wires. In fact it is well to avoid flying in the vicinity of these craft.

Discomfort of Bumps

During their early flying experiences many pupils are apt to become alarmed at the bumps they find in the air. To begin with, these bumps may certainly seem strange and uncomfortable, but, in reality, a pilot with a good machine is absolutely safe in them. In flying, it is a question of the higher the fewer, and quite often a pupil may find that the first 500 ft. or 1000 ft. of atmosphere are bumpy, but at anything above that height the air is quite calm. Bumps have been known to occur at heights of 10,000 ft. or more, but this is very exceptional. It should be remembered that a bump under a wing will only put a machine into such a position as it frequently assumes when turning even at quite moderate banks. Bumps will often be found in the region of clouds, and sometimes, too, they may herald the approach of a thunderstorm.

Sometimes it happens that a pupil, on his first or second solo flight, may lose himself, in which case he has only himself to blame for not studying the lie of the land whilst under dual-control instruction. He should remember, however, that a machine flown solo will rise much more quickly than when two persons are carried, as in dual-control work, and, therefore, it behooves him to keep an eye on the height indicator or aneroid, and see that he does not get higher than he has previously been accustomed to. Again, it is inadvisable for him on his first solo flight to venture into the clouds, where he may easily get lost. He should also remember that, when flying down wind, he will cover the ground very much more quickly than when flying against the wind, which is another fact that may cause him to lose his bearings on his initial flight.

Landing

There is now only the landing to be considered. The pupil knows already that he has to land against the wind, and not facing obstacles, such as woods or hangars. Moreover, he must not land over such obstacles, as they may easily put him off his straight glide to earth, which is the best kind of landing for him to attempt at present. He should know approximately at what height and at what position over the ground he should switch off or throttle down his engine in order to hit off the point on which he intends to land, and on reaching this position he should throttle down his engine fully. On most machines the throttle is set when closed so as to allow the engine to tick over at 200 or 300 revolutions per minute. He must put the nose of the machine down firmly and gradually until the desired gliding angle and speed are attained (this is generally about the average flying speed), and then watch the ground carefully. If there be other machines on the ground which baulk him, it is a wise plan to switch on and to make another circuit while the ground clears. If he overshoots his mark, owing to faulty judgment, or if he sees that he is going to undershoot, it is a good plan to make another circuit, so as to give himself the best possible chance of effecting a good landing. In the latter case he could, of course, switch on his engine, and having gained the necessary distance by flying level in place of gliding, throttle down again, but this may put him off his glide and upset his calculations.

Assuming that he sees that he will hit off the landing ground correctly, he must watch the earth most intently, keeping up his correct glide, angle and speed all the while. When he gets within 20 ft. or 30 ft. of the earth he can begin almost imperceptibly to flatten out by manipulating the control lever in just the same way as that in which he took the machine off the ground a few minutes before. Still watching the ground intently about 20 yds. or 30 yds., or even more, in front of the machine, he continues to pull back the lever ever so gently until he gradually decreases his flying angle, and, finally, as the machine loses its flying speed, its angle of descent is blended in the horizontal of the ground without the slightest jerk, and the machine

White line indicates position of control lever.

If the pilot makes a fast landing and pulls the stick back too quickly in order to get his tail on the ground, he will balloon, as shown in the upper illustration. If he makes a slow landing, he will be far more likely to land safely, even on bad ground. The lower illustration shows what happens when a machine, landing fast, strikes a ridge or uneven part of ground.

comes to earth. In easing back the control lever, it may be necessary at times to pause momentarily, or even to move the stick forward slightly, if the backward movement has been made too suddenly, or too much, and has resulted in the machine being unduly held up, or even made to "balloon." The whole art of landing consists in accurately timing the relative position of the ground and the machine by the eye working in perfect unison with the hand. The reason why the pilot must look ahead to land, and not directly beneath him, can be explained by reference to the fact that in a fast travelling railway train single sleepers on the track can be seen if the observer looks ahead out of the window, whereas if he looks down beneath him the sleepers pass him in a continuous blur. So in landing; the further one looks ahead the slower the ground appears to be travelling, and vice versa.

A pupil must remember from his earliest flying experience that he must put the nose of the machine down if for any reason the engine stops or slows. Generally speaking, he will come to no harm through going too fast in the air, which simply means that he is going downwards; but he will easily get into difficulties if he tries to fly too slowly, which means that he is trying to climb the machine at an angle so great that there is not sufficient speed of air under the wings to maintain them in flight. He must also remember the importance of keeping the machine heading perfectly straight in the direction in which it is intended to land (this direction being determined at a height of several hundred feet above the ground), and of not allowing it to swing out of its course, either owing to bumps or to the necessity of having to use more engine.

Land on a Mark

No pupil can expect to make a perfect landing at the first, nor yet the second attempt. It is all a matter of practise, and the best way to obtain proficiency is to practise landings. He can also practise landing on one particular spot in the aerodrome. By flying in a big aerodrome he is all too apt to allow the machine to land where it wants to, instead of making it land where he wants it. This tendency should be strenuously avoided, and the pupil, especially when he becomes more proficient in the use of "S" turns, should always fix on some definite spot on the ground where he intends to land, and then make his machine land as near that spot as possible. To begin with, he should not make his "S" turns too near the ground, but as he becomes more proficient he can turn lower and lower, always remembering that if he has plenty of speed on the machine he will also have plenty of control over it. The reverse is equally true.

Landing Mishaps

، There are two alternative faults in landing, one very much worse because it is more dangerous than the other.

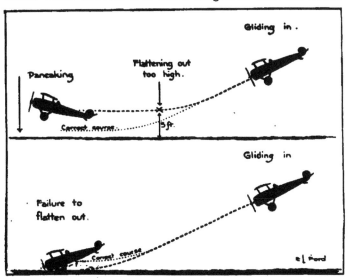

Bad landings. "Pancaking," result of flattening out too high; and the effect of omitting to flatten out, or flattening out too late.

A pupil can either fail to flatten out and thus fly into the ground, in which case he probably smashes his machine in landing by turning it over; or he flattens out too much or too early. In the latter case a "pancake," or, if it be between 10 ft. and 20 ft., a "stall" results. A pancake means that the machine lands several feet up in the air instead of on the ground, and then, having lost its flying speed, it pancakes, or drops to earth. If the pancake is not very pronounced, no damage may be done, as the elastic shock absorbers on the under-carriage will take up a great deal of it.

When the machine has once landed the pilot should allow the tail to drop on to the ground of its own accord. He must not pull it back onto the ground, for if this is done the machine will often leave the ground again for a height

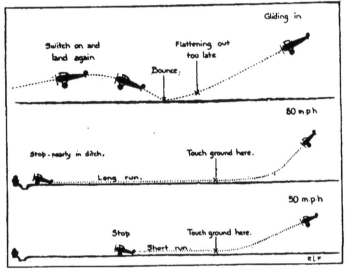

Preventing a bad landing by putting on the engine after bouncing. Fast and slow landings and their ultimate effect on the "carry" of the machine.

of several feet. In some machines the tail is purposely held up off the ground as long as possible.

A worse variety of pancake will result in a broken under-carriage, but no other damage to machine or pupil. The worst form of pancake, which takes place when the machine loses its flying speed at a height of 20 ft. to 30 ft., often develops into a stall, which means that the machine loses its flying speed in mid-air, followed probably by a sideslip, when the machine crashes down on one wing and is hopelessly smashed. In this case the pupil may be hurt, although, if the height is not great, he may not be seriously injured, as the wings act as a cushion to the blow by breaking up and thus absorbing the shock. The only method of saving such a catastrophe is to put the nose of the machine down to allow it to regain its flying speed, or else put on the engine and so allow it to pull the machine out of its precarious position: both these operations could only be performed by experts, as the air space is generally too small to allow of this evolution being performed successfully. An expert would get the machine to earth safely by putting the control lever forward and then moving it back again instantly; but the operation requires a quick touch and an accurate eye if it is to be performed successfully. Pupils should not try such a manœuvre, as it will generally result in their flying into the ground. If the engine has failed, they will do better to allow the machine to pancake.

Bouncing and Bumpy Landings

There is another faulty method of landing, which results in the machine leaping or bouncing over the ground, owing to its speed being too great when it touches the ground first and the angle of descent not being sufficiently small to allow it to run along the ground. The best method of counteracting this fault is to put on the engine slightly between the bumps and then to flatten out and attempt to make a better landing. This also is by no means an easy operation, even for an expert, whose machine may have been made to bump unexpectedly by striking a ridge of ground, or an unseen bank. If the pupil pulls back the control lever too quickly, when the machine still has flying speed, he will "balloon," or go up, and had better then put his engine on and try again after another circuit. On most machines an ideal landing would be to allow the tail skid and the wheels to touch the ground together. This is called a three-point landing and indicates that the machine has been held off the ground up to the very last moment.

In landing, it is always good practice to come down slowly and to attempt to strike the ground at the lowest possible speed in conjunction with safety. This does not mean that it is advisable to glide as slowly as possible, as this is dangerous practise, and, and if turns be attempted on a glide of this kind a stall may result. Instead, it means that when the machine is within 15 ft. to 30 ft. of the ground the speed may be cut down so that the velocity is as low as possible near the ground. The advantage of this is that the run, or the carry, of the machine when it has once landed is considerably reduced, which assists the pilot in making a landing in a small field. If he landed very fast his carry would be further, and there would be more chance of his running into obstacles or turning over if he had to land on soft ground.

Some pupils, when they see that they are going to over-shoot the mark, instead of making another circuit and profiting by their previous mistake, attempt to hit off their landing place by bringing the machine down at a much steeper angle than its ordinary flying speed. This is a common fault and is quite ineffectual, because the machine will have gathered so much extra speed on its descent that it will lose its flying speed much more slowly when flattened out, and will glide across the aerodrome and finally run into a hedge or dyke at the other end owing to its increased momentum.

Taxying

The last point in aerodrome flying, and by no means the least important, is the manipulation of the machine in taxying. This means driving the machine on the ground and steering it with the rudder. Sometimes the ailerons are also worked, but in the opposite direction to that adopted in the air, i. e., the control lever is moved to the left if it is desired to turn the machine to the right; the right aileron comes down and acts as an air brake on that side, so that the machine turns about that wing tip. Taxying should always be done slowly, especially on bad ground, as much damage can be done to a machine by rough or careless handling.

When it is desired to turn the machine on the ground it is sometimes advisable to get up a little speed, raising the tail of the machine by putting the control lever forward

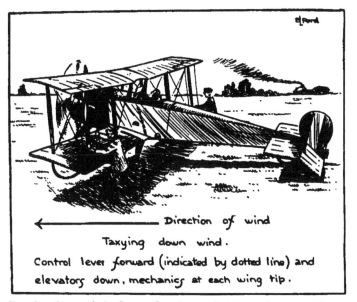

Direction of wind

Taxying down wind.

Control lever forward (indicated by dotted line) and elevators down, mechanics at each wing tip.

Taxying down wind. In taxying down wind, keep the control lever central or a little in front of central, so that the wind cannot get under the elevator and overturn the machine. In landing in a strong wind, wait for mechanics to hold down the wings before taxying back to the shed.

and then ruddering in the desired direction. On some machines the rudder and tail skids are interconnected, and, in consequence, the machine can be steered on the ground with the greatest accuracy and ease. If a machine repeatedly swings sideways when landed or being taxied, the undercarriage is probably out of line with it, and must be trued up. When taxying over heavy or rutty ground it is a good plan to hold the control lever well back so as to keep the tail of the machine on the ground. When taxying across ruts an excessive amount of engine should never be used, for if this is done and the tail of the machine be up, there is an excellent chance of the whole aeroplane falling over on its nose. In windy weather it is sometimes very difficult to control the movements of the machine on the ground, in which case it is always advisable to order two mechanics to hold the wing tip struts and escort the pilot back to the sheds, or out to the starting ground, as the case may be. In taxying down wind keep the stick forward so that the wind cannot get under the elevator, which is down, and so blow the machine on to its nose.

Obtaining a Pilot's Certificate

When the pupil has made one or two successful flights he may wish to qualify for his pilot's certificate, which is issued by the Royal Aero Club of Great Britain and Ireland. His instructor will provide him with the necessary forms, which he has to fill up and forward, together with his photograph and the sum of £1 1s. to the Club whereupon his name will be placed upon the list of pilots, and he will receive his brevet or ticket, which resembles a motorcar license, and is useful to take with him on cross-country flights, where it can always be used as a reference or proof of identity. The address of the secretary is 3, Clifford Street, New Bond Street, London, W. 1.

The test imposed before the certificate is issued is a simple one. The pilot must make two solo flights and execute a certain number of figure-of-eight turns in the air. He must also land with his engine cut off from a height of 100 metres, whilst on the first part of the test he must land within a certain distance of two marks on the ground. It does not follow, however, that because a pupil has taken his ticket he is an expert aviator. The ticket flight is the simplest possible test, and he still has a long way to go before becoming a qualified pilot. The chapters that follow will show that this is so.

CHAPTER VI

The Use and Working of the Instruments

ALTHOUGH it is quite possible to fly well and accurately without using instruments, they are often useful for verifying one's position in the air and finding the way across country.

The instruments generally found on school machines are the compass, air speed indicator, height indicator, sideslip indicator, fore-and-aft level, engine revolution counter, watch, petrol gauge, pulsator glass or sightfeed oil dome, and the pressure gauges for the oil and petrol supply. There is also the map carrier. These instruments and fittings should, so far as possible, be grouped together so that they can be seen by the pilot with as little deflection of his eye from the horizon as possible. Generally, a total weight of about 15 lb. or 20 lb. is allowed for the complete equipment.

Illuminating the Instruments

As the instruments may be used at night, some way of illuminating them must be employed. Generally the indicator fingers and the figures on the dials are painted with luminous paint so that they can be seen in the dark. It is also quite usual to find a small electric-lighting set with battery incorporated in the design of the instrument board, with a switchboard for switching on a hooded light to any or all of the instruments it is desired to read. Each is shielded to prevent the pilot's eyes from being dazzled, and can be switched on independently of the others when desired.

Some notes on the instruments chiefly used on school machines will be of interest. Every endeavour should be made to find out not only how they work and how to use them, but what failures are likely to occur and how they can be most quickly put right.

The Compass

The magnetic compass is a most important addition to any aeroplane intended to be flown across country. It should be placed in full view of the pilot in the fore-and-aft line of the machine when in the flying position. With the tail of the machine raised several feet from the ground, the compass should be set perfectly level, both laterally

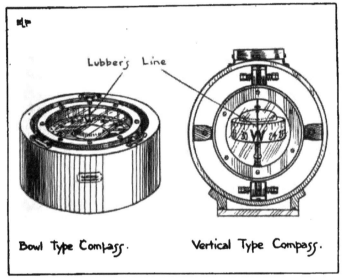

Lubber's Line

Bowl Type Compass.

Vertical Type Compass.

The compass. Two typical types of aeroplane compasses. Left: The bowl pattern. Right: The vertical type, which is read from the back.

and fore and aft. The magnetic compass is an instrument for indicating the magnetic north, as the red end of the magnetized needle to which is attached the compass card always points to the magnetic north, unless affected by magnetic material near it on the ground or carried in the aeroplane. The compass card is pivoted at a point so that it is free to revolve in the horizontal plane. To damp out vibration it is immersed in a mixture of distilled water and

alcohol (two parts alcohol to three parts water) contained
in a kind of bowl, made of some non-magnetic metal,
with a glass top or side to it, through which the pilot can
read the bearing on the card. A mark is painted on the
body of the compass, outside the bowl, in the direct fore-
and-aft line, both of the machine and of the compass. It
is called the "lubbers' line." As the compass card rotates,
caused by turning the machine in the air, the figures on it
register with the lubbers' line, and give the pilot the direction in which the nose of the machine is pointing from
time to time.

The compass card is circular and is divided into 360
degrees. The marking of the card is carried out clock-

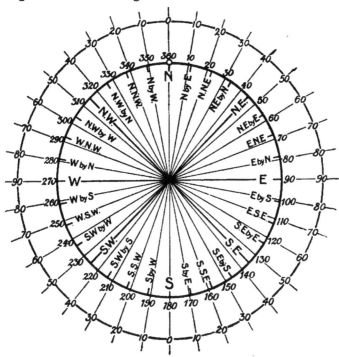

The dial of a compass.

wise, i.e., N. = 360 degrees or o degrees; N.E. = 45
degrees; E. = 90 degrees; S.E. = 135 degrees; S. =
180 degrees; S.W. = 225 degrees; W. = 270 degrees;
N.W. = 315 degrees; and N. = 360 degrees again. The
lubbers' line, as well as the figures on the compass card,
being painted with luminous paint, can be read in the dark.
The cardinal points are N., S., E. and W.; the quadrantal
points are N.E., S.E., S.W., and N.W. It is worth remembering that each degree contains 60 minutes (denoted 60'),
and that each minute contains 60 seconds (denoted 60")
of an arc. One point of the compass is equal to 11° 15' 0",
which means 11 degrees 15 minutes, so that there are 32
points in the compass. (See sketch on page 90.)

Compass Error

If the compass needle is not affected by local magnetic
material in an aeroplane, its red end will point to magnetic
north and its blue end to magnetic south. The angle that
the needle is deflected from true north is equal to the magnetic variation. The angle that the needle is further deflected from magnetic north by magnetic material carried
in the aeroplane is called deviation. The total error of
the compass needle from true north caused by variation and
deviation is called the compass error. Compasses must be
swung from time to time to test their accuracy, as is
explained in Chapter VIII, or if magnetic material is added
to the aeroplane, in the shape of machine guns, bombs, tool-
kits, etc., it will cause the needle to deviate from its correct
angle. This error can be corrected, to some extent, as
explained on page 129, by inserting small magnets in little
holes provided for them in the fore-and-aft line, and also
athwart the compass. On some compasses will be found a
small wooden box fitted to the underside with holes for the
accommodation of correcting magnets, while in others the
slots for the magnets will be found on the top of the compass.

A common trouble with compasses is caused by a bubble
forming in the liquid. To extract this the filler plug must
be removed from the bowl, the bubble brought up into the
plug hole, and distilled water added until the bowl is brim

full. A fountain-pen filler or glass tube which will allow
the bowl to be filled drop by drop may be used for the purpose. To counteract the expansion of the liquid due to
changes in temperature, a small chamber of thin metal, capable of expanding and contracting slightly, is connected
to the compass bowl by a small passage through which the
liquid has access to the expansion chamber. Sometimes a
bubble can be removed by pressing the expansion chamber
when filling the bowl through the plug.

Great care is taken in compasses to insulate them from
vibration. Sometimes the bowl is allowed to rest on a
horsehair pad in a metal cup. It is also provided with rubber or felt shock absorbers, which assist in damping out
the vibration and making the card easy to read.

Compass Spinning in a Cloud

The compass is an extremely reliable instrument, and
pupils should accustom themselves to steering by it and
learning to study its reading. If the compass card begins to
spin when the machine is in a cloud, for instance, it is a sign
that the machine itself is turning and not the compass card,
the north point of which is always trying to point to the
magnetic north as the machine is turning round. Sometimes
the compass card can be used as a lateral level, for unless
it is lying parallel to the horizon it shows that the machine
itself is tilted. In the same way it can be used as a fore-and-
aft level.

In some compasses the card is designed to be read on its
side instead of from above. For this purpose it is attached
to the rim of a flat float inside which the magnetic needle
is carried. The float is pivoted, and the marking of the
compass card, being on its rim, can easily be read as it
changes its position. This type of compass is read from the
back, and the figures omit the final "o" in the bearings for
the sake of clearness. Thus, when the compass card reads
27 against the lubbers' line, which is also at the back of the
instrument, the nose of the machine is pointing 270 degrees.
The card, which is in the form of a vertical band round the
rim of the float, is painted red and blue to represent the
northerly and southerly semi-circles. This type of compass

is used a great deal on modern aircraft, and pupils would do well to familiarize themselves with its design and mechanism, as it is rather different from the more ordinary and older patterns, in which the compass card lies flat and is read from above.

The Air Speed Indicator

The air speed indicator is an instrument used to inform the pilot of his speed through the air and not over the ground. It consists, roughly, of two parts—the instrument and the Pitot tubes. The tubes are fitted to the leading edge of the top wing, generally to one side of the central fore-and-aft line of the machine, so as to be out of the way of the slip stream of the propeller, which would cause faulty readings. On a pusher or a double-engined machine, with the engines placed one on each side of the centre line, the

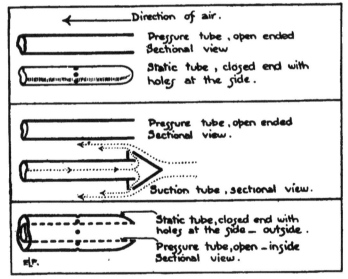

Aeroplane instruments. Pressure and static tubes used in conjunction with the speed indicator. The arrangement is known as the Pitot tube.

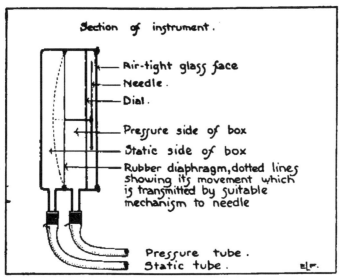

Section of instrument.

— Air-tight glass face
— Needle.
— Dial.

— Pressure side of box
— Static side of box
— Rubber diaphragm, dotted lines
  showing its movement which
  is transmitted by suitable
  mechanism to needle

Pressure tube.
Static tube.   ELF.

Aeroplane instruments. Section of air speed-indicator box, showing the rubber diaphragm and pressure and static sides of the instrument.

tubes could be fitted centrally, as the propeller draught would not affect them. There are two tubes arranged either side by side or concentrically, one inside the other. (See sketch on page 93.) One tube has its end open to the air and the other has its end closed, but is provided with a ring of small holes in its side, or, in the case of the concentric tube, the inside tube has its end open to the air, whilst the outside one, with the holes in it, has its end tapered down and soldered to the inner tube. The first tube registers the pressure of air meeting it and rushing down the pipe leading to the inside of the instrument, which is enclosed in an air-tight case. In this case is a rubber diaphragm, which is compressed by the force of air created by the passage of the open-ended tube through the air when the machine is in motion. This tube is called the pressure tube. The other tube is connected to the underside of the rubber diaphragm which divides the instrument box into two air-tight compartments, and transmits the pressure of the air at rest to this side of the diaphragm; thus a balanced result, or reading, is obtained. The difference in the pressure in the two tubes is proportional to the square of the velocity. Air speed indicators under register with height, and to correct this error a simple method is given. Multiply the reading of the air speed indicator by the reading of the altimeter in the number of thousands of feet and divide by 60. This answer is the correction to be added to the reading of the air speed indicator. Example:

$$\frac{100 \times 6}{60}$$

Air speed indicator registers 100 m.p.h. at 6000 ft.

= 10. Therefore the true air speed at this height is 110 m.p.h.

Faults in Air Speed Indicator

The up-and-down movement of the side of the diaphragm is conveyed by minute and very delicate mechanism to a needle, which works round as a pointer on a dial marked off in miles per hour or knots. Generally, the lowest reading given is about 40 m.p.h. and the highest 160 m.p.h. To secure the correct functioning of the instrument, the following points are of importance:—(1) The tubes should be pointing straight against the air stream when the machine is in the flying level position: if they are canted up or down they will, obviously, give a faulty reading. (2) There must be no air leaks in the connections between the Pitot tube and the instrument. (3) Rubber joints are used in several places, and these should be properly secured by copper wire, not crossed or kinked, and should show no signs of perishing. These rubber joints are used so that the water which might accumulate in the metal tubes may be drained off when desired. (4) The instrument itself must be airtight, and for this reason a rubber packing washer between the glass cover and the body of the instrument must be properly fitted and the two parts screwed well home. Sometimes a careless mechanic may connect up the tubes vice versa to the instrument, or cross them at one of the rubber joints, so that an erratic reading will result. On most instruments the unions for the correct tubes are marked: P=

pressure, or open-ended, tube, and S = static tube. It is
never advisable to blow through the tube, as moisture from
the mouth may get into it. The system can be tested better by sucking through the tube and then closing the open
end and seeing if the needle moves. If it does, there is a
leak in the instrument or tubes.

The Height Indicator

The height indicator is operated by the pressure of the
air, and is affected by variations of temperature. It consists of a dial marked off in hundreds and thousands of feet
from zero to 10,000 ft. or 20,000 ft., depending on the
scale and the type of work for which it is intended. The
dial can be set to zero by turning a milled ring. This would
be necessary after a fall or rise of the barometer, which
would upset the reading of the indicator. It would also be

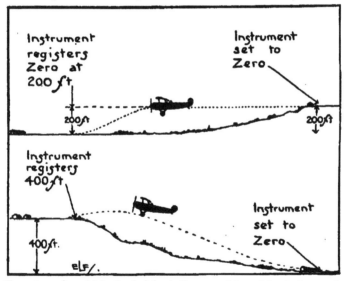

Two cases in which the height indicator might mislead the pupil.
The height indicator only registers the height of the machine above
its starting point, so unless the ground over which the pupil flies
is level with his starting point, the indicator will not be accurate.

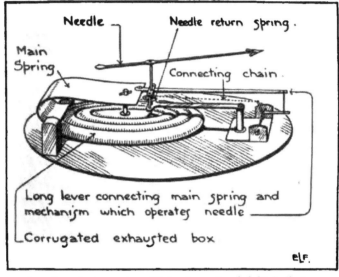

Needle

Needle return spring.

Main Spring

Connecting chain.

Long lever connecting main spring and mechanism which operates needle

Corrugated exhausted box

e.l.f.

Aeroplane instruments. Mechanism of a height indicator.

necessary, in the case of the machine being transferred from one aerodrome to another, when the latter is either higher or lower than the former. The height indicator is always set to register zero at the ground-level where the machine is stationed, and it does not register sea-level unless a special compensation is allowed for between the height of the aerodrome and the mean height of the sea. The height indicator is simply an aneroid barometer. The indicator needle is connected by suitable and very delicate mechanism to an evacuated circular metal drum with corrugated top, which expands as the air pressure diminishes with height and actuates against a spring. It contracts again as the pressure increases, the spring transmitting the movement to the needle. As the metal of which the box is made is very thin, the pressure of the air tends to make it collapse altogether. This is prevented by fastening the bottom of the drum to the bottom or bed of the instrument, and the top side to a spring. As explained, the pressure of the atmosphere in the box compresses it against the spring, from which the movement of the box is transmitted by suitable mechanism to the indicator needle. The dial of the instrument is calibrated correctly in feet, so as to give approximately correct readings of the amount of expansion or contraction of the exhausted drum. With both air speed indicator and height indicator there is a very considerable lag in their action, so that a reading of both instruments must only be taken as approximate at any given time. Height indicators give very little trouble in operation, but it is advisable, when anything goes wrong with them, to send them to an expert for repair rather than to tinker with their delicate internal mechanism.

Contouring a Flight

A barograph, or recording barometer, is sometimes used as a height indicator to show graphically the course taken by a machine in flight. The paper on which the recording needle writes is graduated vertically for this purpose in feet, instead of in millimetres or inches of mercury, and horizontally in hours and minutes. It is situated on a drum driven by clockwork and made to revolve very slowly. If a pilot winds up the clockwork in starting he will be able, at the end of his flight, to see his up and down course through the air, to work out how long it took him to climb to such and such a height, and to note whether his rate of climb and glide was constant. He can also tell at what height he was at any moment of his flight. These recording barographs are sometimes used by pupils undergoing height tests, or when making cross-country flights, and a great deal of useful information can be deduced from the study of the completed chart of a flight. When a large number of these charts have been made, the instructor can use them to judge the skill of various pupils in climbing or gliding. A steady climb to 10,000 feet is indicated by a steady line, but if the line on the chart goes up in jerks or steps it shows that, for some reason or other, the pupil did not climb steadily, or else did not get the best results out of his engine whilst performing the test.

Trouble with barographs may be caused by friction be-

tween the pin and paper, or in the levers actuating the
pointer. If the chart shows a number of quick, jumpy lines,
friction in the mechanism is indicated.

The Sideslip Indicator

The best sideslip indicator is undoubtedly provided by a
simple piece of string or tape fastened to a strut or cross
bracing wire of the machine in view of the pilot. It is allowed to fly back by the forward motion of the machine in
the air. When the machine is flying straight, or being
turned with a correct amount of bank, the string will continue to trail straight back dead in the fore-and-aft line of
the machine. If, however, it tends to trail out sideways,
either one way or the other, it indicates a sideslip in the
direction away from which the string is trailing. Thus, if
the string is trailing out to the left from its anchoring point
on a left-hand turn, it shows that the machine is slipping

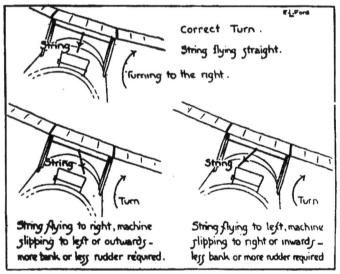

Movements of the string sideslip indicator, which show whether
turns are being made correctly.

to the right or outwards. If, however, it is trailing to the left on a right-hand turn, it indicates an inward sideslip to the right. The pressure of the wind noticed by the pilot on his cheeks should also tell him that he is making a faulty turn. Wind on the right cheek on a right-hand turn indicates an inward sideslip to the right; wind on the left cheek in similar circumstances would indicate an outward sideslip to the left.

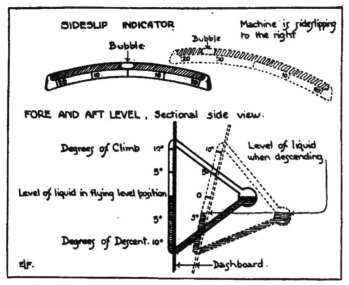

Aeroplane instruments. The sideslip indicator and fore-and-aft level. A non-freezing liquid such as alcohol is used in both instruments.

The cross spirit level is also used as a sideslip indicator. It consists of a bowed tube, filled with a mixture of alcohol and water to prevent it freezing. When the machine is flying horizontally level the bubble is in the centre of the level, which is marked 0°. When a wing drops, the bubble moves away from the low side; but on a turn correctly executed it will remain central, as the centrifugal force, which tends to throw the liquid outwards, will have been balanced by the correct amount of bank. If the bank is too small for the turn, the bubble will travel to the inside of the turn as the centrifugal force will cause the liquid to fly outwards. The remedy in this case would be more bank or less rudder. If the bank is too great, the bubble will go to the outside, while the liquid slips inwards. The cure is to put on more rudder or to decrease the bank. These remarks apply to all moderate turns, banked up to 45 degrees, and for all turns correctly executed, however steep, the bubble should remain central in the cross level. The use of this instrument in the air is much more likely to muddle beginners than to be of assistance to them. It is far better for them to learn to fly accurately by sight and feel or touch of the machine than to fly by the use of instruments, which, although useful as a means for checking the accuracy of their judgment in flying, are liable to mislead them and to break down.

Horizontal Sighting String

Some pupils have been materially assisted in their early flying by fitting a string stretched horizontally level across the machine dead in their line of sight. This horizontal sighting string, when compared with the horizon, assists a pupil to maintain his machine laterally level. Unless the sighting string is parallel with the horizon, the pupil knows that he is not flying laterally level. If the string dips below the horizon to the left, he is obviously flying with his left wing down, and he must move the control lever to the right to pull the wing up, centring it again as the machine comes up level. Assuming that when the machine is flying level in the fore-and-aft line the string is set to be dead on the horizon, from the pupil's point of view, the pupil can see in a moment if he allows the machine to get its nose down or to climb. In the first case the string will drop below the horizon, whilst in the second it will appear to rise above it. It has been found a great help to the pupil to have the sighting string dead in line with his eye, as otherwise he would have to judge his position laterally and fore and aft by comparing the nose of his machine and the underside of the top plane with the horizon. This double comparison

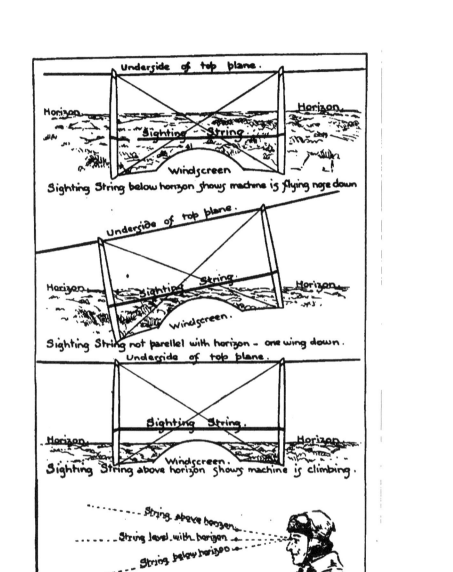

Underside of top plane.

Horizon. Horizon.

Sighting String.

Windscreen

Sighting String below horizon shows machine is flying nose down

Underside of top plane.

Horizon. Horizon.

Sighting String.

Windscreen.

Sighting String not parallel with horizon — one wing down.

Underside of top plane.

Sighting String.

Horizon. Horizon.

Windscreen.

Sighting String above horizon shows machine is climbing.

String above horizon

String level with horizon

String below horizon

Line of sight indicated by dotted lines

Elford.

The uses of the sighting string for determining if the machine is flying horizontally or laterally level, if the machine is climbing, flying level or descending. For preference, the string should be set to be dead in line with the pupil's eye and the horizon when the machine is flying level. 102

is too much for the beginner, who finds it difficult to gain any idea as to the position which the machine is assuming in the air, whereas with the string he·has only one point to keep in mind. The fitting of the string also teaches him to look out ahead of his machine, which is the proper direction in which to concentrate, instead of looking at instruments or aimlessly watching odd pieces of ground, sky or machine that may attract his attention from time to time, which, until he becomes very much more proficient, cannot possibly assist him to keep the trim of his machine correct.

The Fore-and-Aft Level

The fore-and-aft spirit level consists of a triangular tube with a bulb in the apex of the triangle, which is set at the back of the instrument board. The front of the level is graded in degrees and lies flush in the dashboard, the liquid being arranged to register 0° in the half-way position when the machine is flying level. When the machine is climbing, the liquid rises in the glass, and when the machine descends it recedes. This level is only useful in so far that it tells the pilot the degree of climbing angle of which his machine is capable under any particular set of conditions of engine power or load. Unlike the air speed indicator, which will always allow him to know if he is flying within safe limits (it does not matter whether his engine is on or off, or what the load is), the fore-and-aft level cannot be used for this purpose, as a machine might be able to climb at a certain angle with the engine running all out or with a light load, but it could not reproduce this performance with a failing engine or full load. Thus the pupil should not depend upon the fore-and-aft level too much as a means for telling him his position in the air, i.e., if he is climbing, flying level, or descending. If he noticed that he was flying level under any particular set of conditions with the fore-and-aft level registering at a certain mark, he might attempt to keep the level at the same mark if he ran into a big cloud unexpectedly. Even then his air speed indicator would probably give him a better idea of his fore-and-aft position than the fore-and-aft level itself.

The Engine Revolution Counter

The engine revolution counter can be driven off the crank-shaft, the camshaft, or the pump shaft of the engine. Providing that a correct gear ratio is interposed between the shaft on the engine and the flexible shaft of the indicator, it will indicate the actual number of revolutions of the engine per minute. Like other instruments, the needle and figures on the dial are often treated with luminous paint so that they can be read in the dark. The dial is arranged according to the speed at which the engine runs. It may be graduated from 500 r.p.m. to 1500 r.p.m., or up to 2000 or more r.p.m. if the engine is a high-speed one. The chief troubles with revolution counters occur owing to the break-age of the flexible shaft, or the pin coupling at the engine, pump or countershaft connection breaking or coming adrift. Sometimes oil will work its way up the flexible shaft and cable and enter the instrument, where it will cause the hand to stick. The cure is to clean both the flexible shaft, the cable and the instrument with petrol. Revolution counters work either by centrifugal force or electrically. If the internal mechanism breaks down, the instrument must be returned to an expert for repair. In the event of any of these instruments failing in the air, there is no immediate need for the pupil to land.

The Watch

A watch is always a useful instrument in an aeroplane. It should have illuminated fingers and figures and be well insulated from vibration. This is done by placing it in a felt-lined case. A watch being valuable not only to aviators, it is advisable for it to be secured to the machine, or else to remove it when not required, as in the case of a machine being left out in a field all night, for otherwise, when the pilot returns, he may discover that some one had appreciated the value of the instrument and appropriated it for his own use.

A map carrier can consist either of a leather case with a celluloid cover through which the map can be read, or else of a flat tin or sheet aluminium box with a roller at each side. In the latter case, clips are arranged to grip the map in each roller. The map must be cut to the correct width so that the carrier can accommodate it, and is then wound on to one roller and secured to the other. By turning the roller, the map is gradually unwound underneath the celluloid cover, which has to be removed to fit the map into the case, and stretch by stretch becomes visible to the pilot's eye. The

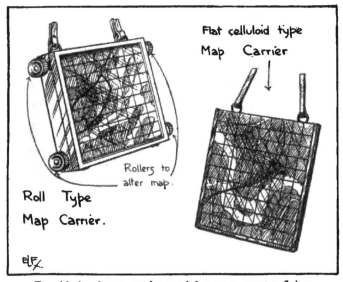

Two kinds of map carrier used for cross-country flying.

principle on which this type of map carrier works is similar to the mechanism for winding up a Kodak film. It is chiefly useful when a long stretch of country has to be flown over where the course to be travelled is too great to be viewed easily on one map, and the one sheet may not cover the one journey. Several map sheets can then be gummed together and cut so as to form one continuous stretch. The map case is carried slung round the neck by a piece of tape, but the pupil must be careful to see that it does not foul his controls, as it may do if it were left to dangle freely, or if it were hung up on the machine.

Petrol and Oil Gauges

The petrol gauge consists of a vertical glass tube connected to the petrol tank by short lengths of pipe at the top and bottom. A tap can be fitted between the gauge and the tank, in which case the gauge will only register when the tap is turned on. The petrol takes up the same level in the gauge as in the tank, so that the pilot can see exactly how much petrol remains. He should remember, however, that, as the gauge will probably be fitted in the back of the tank, the reading will only be accurate when the machine is flying level. When the machine is climbing steeply the petrol will mount up at the back of the tank and in the petrol gauge, giving a reading in excess of the amount carried. When the machine is descending the reverse occurs.

Fitting Nuts and Bolts

A point in connection with the arrangements of taps worth noting here is that they should always be designed to hang down in their running position. There will then be no chance of them closing themselves unexpectedly owing to vibration in the air. In the same way, the bolts should always be put in with their securing nut underneath, so that, in the event of the nut falling off, the bolt will still remain in position. In cases where bolts are put in horizontally, and not vertically, the head should always face the direction of flight where possible, so that, should the nut come off, the tendency of the bolt will be to remain in its socket.

The sightfeed oil pulsator domes, used on most rotary engines of the Gnome type, are connected to the oil pipe between the pump and the crankshaft. As the pump discharges a small amount of oil at each stroke of the engine, the force of the pulsation is also transferred to the oil in the oil pulsator dome. The oil rises slightly and then falls again. If the gear ratio between the pump and the engine is known, it is possible to calculate the number of engine revolutions by timing the number of pulsations per minute in the glass dome. In the Gnome engine the pump shaft turns at $7/4$ of the speed of the engine, while in the Renault engine the pump shaft turns at $2/7$ of the speed of the

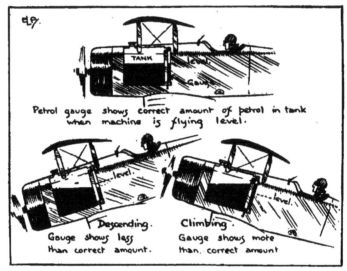

Petrol and oil levels, showing why the petrol and oil gauge fitted
to the back of a tank may mislead the pilot as to the amount of fuel
remaining, unless he is flying level when he reads the gauge.

crank-shaft. There is no pulsator gauge in the Renault or
R.A.F. engines, which are commonly used in school machines. An ordinary pressure gauge is often fitted instead.
When the oil is cold and the engine just started up the
pressure will be seen to rise, but after the engine has been
running for seven or eight minutes it will drop gradually to
its normal mark.

Pressure Feed

A similar gauge is used in connection with pressure-fed
petrol tanks, into which air is pumped, either by a hand
pump or else mechanically by the engine, or, again, by a
fan operated by the passage of the machine through the air.
Generally a pressure of 2 lb. or 3 lb. per square inch is sufficient, and should the pressure increase beyond normal,
some kind of adjustable safety valve or tap should come

into action to prevent the tank being unduly strained. Pupils are often liable to forget the necessity for keeping pressure in the petrol tank by hand pumping from time to time, and have suffered forced landings in consequence. The best type is that in which the hand pump is only used as a stand-by or for starting up, the pressure being mechanically maintained as soon as the engine is running. Leakage of pressure may be due to the petrol filler cap not being properly screwed down, or to taps or safety valves not seating properly, in which case they must be ground in until an airtight joint is obtained. Failure to hold pressure may even be due to a careless mechanic fitting the oil tank filler cap, which generally has a hole in it, to the petrol tank and vice versa when last replenishing the machine.

There are many other kinds of instruments to be found on Service machines, but, beyond mentioning these, there is no need for the beginner to worry his head about them. Such instruments as a petrol flow indicator, radiator temperature indicator, bomb sights, aeroplane cameras, and aircraft course and distance indicator, are all used on advanced Service machines; but by the time that the pupil is able to fly a machine of this type he will find no difficulty in understanding or operating these fittings.

One Instrument to Check Another

From the knowledge of the possibilities of the instruments a pilot can very often use one as a check upon the other, or use two in conjunction in place of another which may have failed. For instance, in a cloud, if the compass card begins to spin, it is an indication that the machine is turning in the opposite direction, whereupon the pilot should know how to counteract it. By watching the compass card dip or rise, fore and aft or sideways, the pilot knows that his machine is doing just the opposite, and, accordingly, counteracts the movement with his controls. By observing the height indicator he knows if his engine is pulling properly, for if he were to keep the machine at its normal flying speed and it were to drop, as shown on the height indicator,

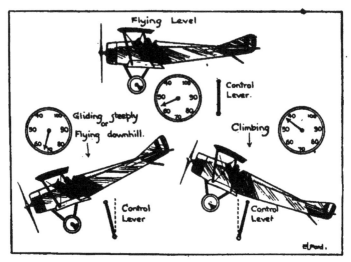

One instrument to check another. Showing how the air speed indicator can be used as a fore-and-aft level. If the machine flies level at 57 m.p.h., it must be descending at all speeds above that, and climbing at all speeds below until the stalling point is reached.

this would be direct evidence that the engine was not giving its full power. The air speed indicator tells the pilot when he is climbing or descending, and also, when used in conjunction with the height indicator, it keeps him informed as to how the engine is running. The engine revolution counter, in addition to indicating the revolutions of the engine, can be used to show whether the machine is climbing or descending as the revolutions slow down or increase accordingly, within certain narrow limits.

The sideslip indicator may either be an instrument based on the principle of having an air speed indicator at each wing tip to register the difference in speed between the two wings, or it may take the simpler form of a piece of string allowed to trail backwards in the view of the pilot. In a cloud the cross level may be used to some extent to maintain the lateral level, the pilot following the bubble with the stick.

Watching the Bubble

The fore-and-aft level could be used if the air speed indicator failed, as it might do through intense cold or condensation and frost. Assuming that a pilot knows at what number of degrees it is safe to climb the machine, a fore-and-aft level can be used successfully for flying, so long as the engine gives its normal power. The side-to-side level indicator shows, within limits, when one wing is up and the other down. The general rule is for the pilot to follow the movement· of the bubble with the control lever. If the bubble rises to the left side it indicates that the right wing is down, therefore the pilot moves the control lever to the left, i.e., he follows the bubble in order to counteract the drooping wing. This rule only applies to small deviations from the horizontal. The bubble is also used to indicate a sideslip on turns. On a perfectly-executed turn the bubble should remain central. If it moves towards the lower wing on a turn it shows that centrifugal force is flinging the liquid towards the outer wing—in other words, that the bank is too small and the machine is sideslipping outwards. If the bubble is near the top wing on a turn it indicates that the machine is sideslipping inwards and that the bank is too steep.

The Importance of the Watch

The watch serves more than one purpose. With it a pupil can tell, more or less, his whereabouts, i.e., providing he has calculated his speed correctly. He can easily verify this by timing himself over the first division of the course which he has marked out on his map. He knows, too, when his petrol and oil supply may be expected to run out, how much daylight he has (an important factor if he starts on a cross-country flight on a winter's afternoon), and in an emergency he can often use his watch to tell him the north, in the case of his compass failing. This is done by pointing the hour hand to the sun, and by dividing the angle between it and 12 o'clock, the result, in the northern hemisphere, being that the dividing line points to the south. If he lost himself and his compass were to fail, he would come down to find his whereabouts on the map with the aid of his watch, and then note some visible and prominent landmark on his right course. He would then fly to it by sight; but, before reaching it, he should pick up two other objects on his correct line of flight, which will help him to keep his course when he has left his first landmark behind. By repeating this process it is possible for him to arrive at his destination without the use of a compass.

Testing the Speed Indicator

The object of describing how one or two instruments can be combined to make a substitute for another is that, when one fails, the pilot need not feel lost. The air speed indicator may fail through a rubber joint breaking or a leak occurring, or even through it being frozen up. This can be tested by blowing down the tube until the pointer shows 70

Importance of a watch. How to tell the north or south in the northern hemisphere by using the watch.

or 80 miles per hour: the tube is then sealed up by holding the tongue over it, and if there are no leaks the pointer will remain at the set mark, falling back to zero when the pressure is released. The engine revolution counter may suffer defect either through the oil being frozen or the drive breaking, while the height indicator sometimes fails or sticks.

CHAPTER VII

Map Reading

IT is necessary for a pupil to be able to read a map before his aerodrome flights can be expanded to a cross-country journey. Those who have studied maps for motoring or cycling before taking up aviation will find that map reading is quite a simple matter. It is necessary to become familiar with some of the more common terms used in map reading and map making. The following have been selected as being of value :—

BEARING.—A true bearing is an angle which a plane or line makes with the true meridian or north. A magnetic bearing is an angle a plane or line makes with a magnetic meridian or north.

A CONTOUR is an imaginary line running along the surface of the ground at the same height all the way round. Each contour represents a fixed rise or fall of so many feet from those next to it. This fixed rise is called the vertical interval, or, for short, "V.I."

THE HORIZONTAL EQUIVALENT, or, for short, "H.E.," is the distance in plan between two contours.

A GRADIENT is a slope expressed as a fraction. Thus a gradient of 1-10th indicates a rise or fall of 1 ft. in 10, or 1 metre in 10 metres, etc.

A DATUM LEVEL is an assumed level, by reference to which heights are measured or compared.

HACHURES are short, disconnected strokes of the pen by which the shading of hilly features is effected. The strokes are drawn directly down the slopes. The closer together the hachures the steeper the slope.

A MERIDIAN is a true north and south plane or line, and a magnetic meridian is a magnetic north or south plane or line.

How the true north of a map
is indicated.

Having mastered these terms, the first thing to do on looking at a map is to find the north. In cases where it is not marked, the top of the map may be taken as the true north, and the borders at the sides as east and west. In other cases, however, two lines will be found, one ending with an arrow indicating magnetic north and the other ending in a star indicating true north. Magnetic meridians indicate magnetic north and can be used accordingly. Meridians on maps are generally true, and not magnetic.

Reckoning the Scale

The next most important point to look at is the scale of the map, for, without a scale, a map might depict a whole country hundreds of miles across, or merely a county or department. There are two methods of indicating the scale: firstly, by drawing one in inches, in which case it is stated on the map how many miles of ground are equal to 1 in. on the map; the second method is indicated by the words, "Representative fraction equals 1-100,000," or some other fraction. In the first case, assuming the scale is 2 miles to an inch. it means that 2 miles on the ground are equal to 1 in. on the map. Thus, by measuring the distance in inches between any two points on the map, the pilot can at once find out how much this represents in miles on the

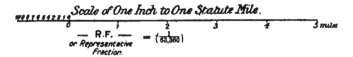

A typical method· of indicating the scale of aviation maps.

ground. Sometimes the letters "R.F." are used instead of the words "Representative fraction." The method of reckoning the scale by means of the R.F. is quite simple. Assuming a scale of 1-100,000, this means that one unit on the map—it does not matter whether it is an inch, a millimetre, or any other unit—is equal to 100,000 inches, or millimetres, or any other unit which has been taken on the ground. The pilot can then work out the scale accordingly. If he works in inches in this particular case, he will find that 1 in. on the map is approximately equal to 1 mile 1017 yds., or, if he works in millimetres, that 1 mm. on the map is equal to 1 kilom. on the ground. Incidentally, this will show the benefit of using the metric system.

Maps Without Scales

It is possible, in certain circumstances, for a pilot to be faced with a map on which the scale is not marked. Obviously, before proceeding to use the map he must discover the scale. He can do this by finding two places he knows on the ground and recognizing them on the map. They may be two villages, a farmhouse and a clump of trees, or two towns. Now, he must find out by measuring—by milestone, or by hearsay, if necessary—the distance between these places on the ground. It can be assumed that it is 4 miles: he then measures the distance between these two spots as shown on the map and finds that it is 2 in. The scale of the map in this case is therefore half-an-inch to a mile.

How Contours are Indicated

Having mastered the scale, the next thing to do is to study the methods employed on a map for illustrating graphically the features on the ground. He will find that heights are indicated on different maps in different ways. Sometimes this is done by colouring, the darkest portions of the map being the highest, and the lightest the lowest. Generally there is a column of colour at one side of the map, and the heights corresponding with the colours marked. The commoner method of indicating height is by contour lines. These can be seen on most maps in the form of irregular curved marks all over the map. A certain set height, called the "Vertical interval," is fixed as a height between these contours, and the amount of the vertical interval is sometimes indicated at the foot of the map. It may be 10 ft., 50 ft., 100 ft., or 500 ft., according to the map. All ground at, say, 10 ft. above sea-level is then joined up by a line on

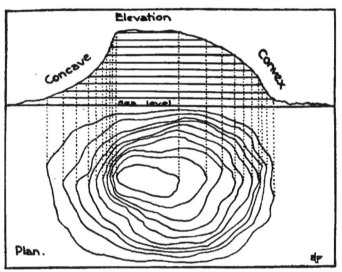

An elevation of contour lines indicated on the map, showing how a concave or convex form of hill can be recognized.

the map, and, assuming the vertical interval is 10 ft., all ground of 20 ft., 30 ft., or 40 ft., and so on, is joined by means of lines, which result in the thin and wavy marks on such maps. Naturally, it is impossible for one contour line to cut another, for this would mean that one part of the ground overhung another considerably. If a pilot knows the height indicated by any one contour line, and also the vertical interval used, he can calculate, by counting the lines above or below and multiplying the result by the vertical interval (which is always given in feet), how much any other piece of ground is above or below the one he started with.

He must remember that on some foreign maps the height is not marked in feet as in English maps, but, perhaps, in metres instead. When studying his map before making a cross-country flight, it is advisable for him to note the highest ground across which he has to fly, and also to caiculate how high it is above his starting point. If, for instance, the aerodrome he started from is 600 ft. above sea-level, and there is ground 2600 ft. high ahead of him, he will know that when his altimeter, which he has set to zero at starting, shows 2200 in the neighbourhood of this range of hills, there is some chance of his hitting them. This might easily happen were he to fly into thick mist or fog during a long cross-country flight; hence the importance of studying the heights or contour levels on a map before starting.

Convex and Concave Configuration

In certain circumstances it may be difficult to find out whether the contour lines indicate a dip or knoll; but, assuming that there is a river or sea coast on the map, the rise of the land can be calculated with this as a base; the flow of a river being known, the lines crossing towards the mouth must be lower than those nearer the source. Generally, however, the heights of contours are marked. Sometimes what is known as a spot level—a small triangle with a height appended to it—will be found at various points on the map. In passing, it may be noted that when the contour lines are close together the slope is steep, and when they are far apart the gradient is more gentle. The hill in which the lower contour lines are far apart and the higher ones close together is said to possess a concave slope; whereas, in a case in which the top contour lines are wide apart and the lower ones close together, the slope is what is known as convex. In a concave slope, the whole of the ground from the top to the bottom is visible from either point; but in a convex slope those stationed at the top and the bottom cannot see each other owing to the bulge in the side of the hill intervening.

A further method of indicating height is by means of hachuring, which consists of a number of short lines drawn down the slopes of a hill.

How Features are Indicated

The colouring of a map is useful in other ways besides indicating height. Woods, for instance, are often coloured green, railways are indicated in black, roads in red or brown, and water in blue. Towns are often shaded in black, although, of course, this is only done in large-scale maps. It is necessary to familiarise oneself with these methods of indicating the main features and landmarks of the country, which will be of the utmost importance when making a first cross-country journey.

There is still room for very great improvement in avia-

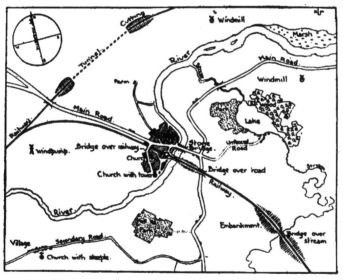

How to read the conventional signs on a map.

tion maps. Motor maps are of little use, as they generally show a confused mass of small details, which are quite un-necessary to the aviator. All he wants are a few general landmarks depicted clearly and accurately, such as large woods, trunk railways, main roads, rivers, lakes and towns. In his work no one of these features alone gives him his bearings, but, rather, their relationship to each other. At a fair height, say, from 5000 ft. to 10,000 ft., many of the small details given in ordinary maps are unnoticeable.

Maps of the Future

Special aviation maps are being produced from time to time in which only main features that can be recognised by an aviator from a height are included. These maps, of course, are on a much smaller scale than the ordinary road map and present the advantage of including in one sheet quite large tracts of country, so that there is no need for the pilot to join several maps together in order to obtain a continuous plan of his course. When a pilot has flown a good deal over certain tracts of country, he will be able to pick out for himself any useful landmark which he can easily recognise. A peculiarly-shaped field, a row of houses, a bridge across a reservoir, a château or monastery with a curiously-shaped entrance drive, or a park placed in an iso-lated position, may give him clues as to his whereabouts which it would be impossible to mark on a map. There is certain to be very great improvement in aerial maps in the future, and it seems probable that photography may be called to the aid of the aerial map-maker.

Studying the Ground Round the Aerodrome

In a succeeding chapter on cross-country flight, such questions as studying the map, working out the course and finding the way when lost will be investigated; but it should be distinctly understood that a pupil cannot put in too much time studying maps before he is actually to use them in flying across country. It is a good plan to obtain a map of the country surrounding the aerodrome and to study the ground from the air, and then compare it with the map. This can easily and most profitably be done during the early solo trips in the vicinity of the aerodrome. Then, when the crucial test of map reading—a cross-country flight—is undertaken, the pupil will not find that flying by map and compass comes strange and difficult to him.

Some countries are more naturally adapted to flying over than others, insomuch as they are more open and have fewer minor features to distract and muddle the aerial map reader. Certain parts of France have this advantage, and in view of the fact that many pupils may make their first solos over French soil, it is necessary to bear in mind that the scale of French maps is given in kilometres, and that the heights are recorded in metres. Eight kilometres are equal, roughly, to 5 miles; i.e., a kilometre is, approximately, ⅝ of a mile. A metre is 39 ins., or rather more than a yard. A hill marked 1000 metres on a French map is therefore, approximately, 3300 ft. high, and not 1000 ft.

Explanation of Map Symbols

The conventional signs used can be found at the foot of most maps. Churches are indicated by a black cross; if they have a tower, the cross is placed on a black square; if a steeple, on a black circle. A double railway line is shown by two thick black lines with intervals of wide spaces in the form of a ladder with very deep rungs. A single line is indicated by a thin black mark with small cross marks cutting it at frequent intervals. Woods are often shown in green with marks to indicate trees. Unfenced roads are shown by dotted lines, and fenced roads by two thin parallel lines. Railway tunnels can be found by noting where the railway line disappears and appears again an inch or so further on, according to the scale of the map. Sometimes the railway is dotted in where it goes underground. Bridges, whether over or under roads, rivers, or railways, are marked with a distinctive marking. Telegraph wires are sometimes shown by a series of dots and dashes. A railway embankment or cutting is shown by shaded lines arranged in a semi-circle about the cutting or embankment and at right angles

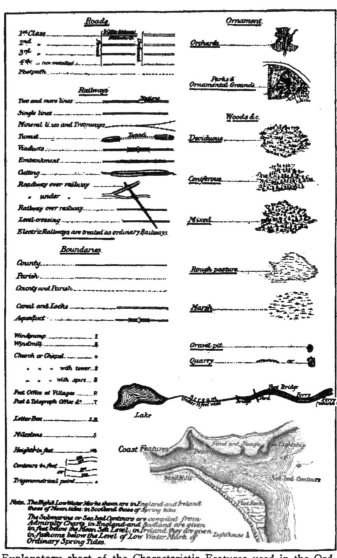

Explanatory chart of the Characteristic Features used in the Ordnance Survey maps.

121

A reproduction from the one-inch Ordnance Survey printed in
colours, which should be compared with the chart of the Characteristic Features.

122

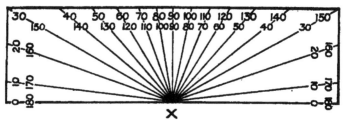

Angles are marked off from the point X, which is placed at the point of departure on the map, the instrument being laid N. and S. from o to o.

to the line. In the case of a cutting, the outer sides of the lines are terminated by semi-circles. (See plate on page 121.)

The Protractor

An instrument that is used considerably in flying, although it is not operated by the pilot in the air, is called the protractor. There are many different models obtainable. Pupils can select one that suits them best. A protractor is a flat piece of wood or transparent celluloid marked off in degrees and in various scales and measurements for use with maps or charts. These measurements may either be in inches or millimetres, or both. A popular type of protractor is illustrated, the method of reading angles or bearings from it being shown. On the reverse side of the instrument will be found a method for reading off the smallest measurement in decimal points. A scale is drawn out in the manner shown. (See sketch.) One of the main divisions is subdivided into ten, and lines are drawn at an angle across it

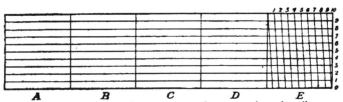

Illustrating the use of the protractor in measuring the distance between two points to ten places of decimals.

as well as parallel with the whole scale, the horizontal lines also dividing the scale into ten. To measure any distance exactly, it is taken to the nearest main division of the ruler, say C (see sketch), and the extra piece which is less than one of the main divisions is measured off against the reduced scale (E). If it measures over 6, but under 7, the exact amount can be calibrated by measuring along down the sloping lines. If the exact distance came to 6 on the top row of figures, plus 5 on the vertical row, the measurement would total ABC + .65 of E, so that, taking the length (A) as one unit, any distance can be measured in terms of that unit to two places of decimals.

Another type of protractor consists of a square piece of transparent celluloid having a circle marked off in degrees from 0 to 360 described about its centre, attached to which is a piece of silk thread. At the top and bottom scales in metres and yards are marked off in convenient scales used in Continental maps, i.e., 1/80,000 and 1/100,000. To take a true bearing with this type of protractor, the central point is placed on the spot from which the bearing is to be taken, with the N. and S. line of the protractor corresponding to the true N. and S. of the map. The silken thread leading from the centre is stretched across the protractor and map to the point the bearing of which it is desired to take, and the angle it makes with the true N. and S. line is read off on the arc of the circle. To lay off a course with this protractor, which is graduated in the same way as an aeroplane compass, the centre point is placed over the starting position and the silken thread stretched across the map and protractor, the arc of which it must cut at the desired angle. A pencil line made down the thread will give the desired direction.

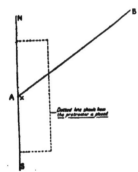

How to lay the protractor on a map.

CHAPTER VIII

Preparing for a Cross-country Flight

THERE are certain pilots who, when told to fly some-where, hurry away into the air without proper pre-liminary study of their course and the country over which they are to fly. No wonder such aviators often lose themselves and meet with other mishaps. It may almost be said that careful preparation for a cross-country flight is half the flight finished.

When a pupil is told that he is required to fly to some spot—which may be called "A"—from his aerodrome at "B," he should first proceed to obtain a suitable map, with the main features clearly marked on it, if possible having the whole journey on one sheet, with a scale of 4 miles to 6 miles to an inch. If his course lies across two adjacent sheets of a map, he must join them together with gum or glue and cut them to fit his roll map carrier.

Studying the Map

He should spread the map out in front of him on a table, examine it carefully, and decide upon the route he is going to take, which may not be in a direct line, owing to the in-tervention of large towns or defended areas. With a ruler he joins the points decided upon, and then, from the scale of the map, works out the distance of the flight. Probably, to begin with, it will not be more than 50 miles or 60 miles.

Now, although it is possible to fly from place to place with the aid of a map alone, a pupil-pilot must remember that in his compass he has a more valuable friend than even his map. If he steers by compass alone and has calculated his course correctly, then, allowing for the time occupied and the speed of the machine over the ground, he can ar-rive at his destination without the use of the map. It is possible for a map to be blown away in the air, or for it to be difficult to read, owing to bumps, and the attention of the pilot being taken up by the manipulation of the machine. When it is remembered that a very great area of land can be seen from an aeroplane at a height of even a few thousand feet, the necessity of steering a course to a degree is not so important as on a ship; moreover, bumps in the air make this almost impossible.

Variation and Deviation

Having been through a course of navigation, he should have little difficulty accordingly in working out a compass course by drawing a picture and, starting with true north,

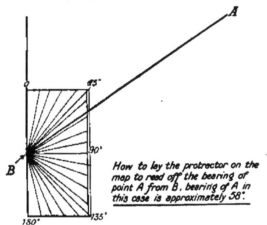

How to lay the protractor on the map to read off the bearing of point A from B, bearing of A in this case is approximately 58°.

How to place the protractor on the map on a true N. or S. line in order to read the true course.

adding or subtracting variation, which, if not marked on the map, must be obtained from a variation chart. This gives him magnetic north, and then, by applying deviation (which is due to the influence on the compass of local iron and steel of the machine), he obtains compass north, and can calculate what the compass will read when he is steering on the true bearing he has worked out from his map.

For instance, let it be assumed that the pupil wishes to fly from B to A, two points on a map. He joins them by a line and then at B draws a true N. and S. line, which he can find, generally, by taking the side of the map as a guide as to the true N. and S., or else the true N. and S. may be marked on the compass rose on the map. He then measures the angles between the two lines, which will give him his true course. He does this by placing the centre of the protractor, on which angles are marked, on the point B in line with the true N. and S. line he has drawn, and reading off the angle that the line B—A makes with it. Imagine it is 58°. He should not have made any mistake either in placing his protractor correctly on point B, or in reading off the angle, because, at first glance, it is obvious that he wishes to fly rather more than N.E. of his present position at B, and he already knows that the compass is marked in a clockwise direction, from O, or 360°, at N., to 90° at E., to 180° at S., to 270° at W., and back to O, or 360°, at N. again. Therefore he knows that the bearing he desires must be a little more than 45°. He then makes a picture of his true course, so as to apply his variation and deviation correctly. Imagine the variation to be 15° W. He draws a line on the picture 15° W. of north, and marks it north magnetic. He finds from the deviation table on his machine that at E the deviation is 3° W., so he draws another line from B 3° W. of magnetic north, as deviation is always calculated from magnetic north, just as magnetic north is calculated from true north. All he has to do now to find his compass course is to measure the angle EBA. The picture has explained to him the following :—

The angle CBA = his true course = 58°.
The angle DBA = his magnetic course 58° + 15° = 73°.
The angle EBA = his compass course 73° + 3° = 76°.

By drawing a picture he can see whether he has to add or subtract his variation or deviation to or from his true course.

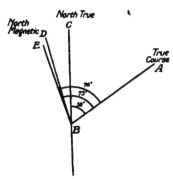

A diagram of a compass course with the variation and deviation applied correctly.

The angle DBC = variation = 15° W.

The angle EBD = deviation = 3° W.

Of course, if the variation were 15° E. and the deviation 3° E. instead, his picture would look like this. (See lower illustration.)

The angle CBA = his true course = 58°.

The angle DBA = his magnetic course = 58° — 15° = 43°.

The angle EBA = his compass course = 43° — 3° = 40°.

The deviation to be applied will be found in the deviation table, which should be pinned up in the machine he is to fly. Regulations are laid down that compasses must be swung or tested for accuracy after a machine has been wrecked or reassembled, after any new fittings have been added to it, when the position of the compass is changed, or when armament, such as bombs or a Lewis gun, have been fitted. If there is no deviation table, the pupil will be well advised to have his compass swung before he leaves. At the same time, so much country can be seen from an aeroplane that a small error in steering is not of vital importance.

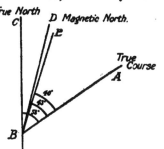

Indicating whether variation or deviation has to be added or subtracted to the true course.

Swinging a Compass

The following is the best method of swinging a compass. At most aerodromes the pupil will find a cement slab with eight lines radiating from the centre, or lines plotted in the ground. These lines indicate the four magnetic cardinal points of the compass, i.e., north, south, east and west, and the four magnetic quadrantal points, north-east, south-east, south-west and north-west. These lines have been correctly arranged by a compass expert with the aid of a landing compass, and wi" be situated far enough away from metal sheds for the metal not to have any influence on the compass.

The machine is then trestled up in the flying position, with its fore-and-aft line laid along the north and south lines on the ground. The pilot, or whoever is swinging the compass, can line up the machine by dropping a plumb line from the centre of the propeller and sighting along it to the tail skid, or another plumb line dropped from the centre of the fuselage until these lines coincide with the north and south line on the ground. It must also be made certain that the machine is dead level horizontally, i.e., from wing tip to wing tip. The lubber's line of the compass should be fitted in the fore-and-aft line of the machine. The compass reading is then taken, and it may be found that, owing to the influence of metal in the machine, it does not read magnetic north as it should do. Deviation will always be east or west, and the pilot must then insert a small field magnet in a slot provided for the purpose, generally in a box under the compass athwartships. If the reading of the compass is less than the magnetic, the direction is easterly, or if the reading of the compass is more than the magnetic, the deviation is westerly (or —). Thus if the machine is placed to head E. by the line on the ground and the compass reads 95 degrees, the deviation is W. (or —). If the deviation is westerly, i.e., the red end of the compass needle is swinging to the left, he must insert the red end of the field magnet to the left, so as to drive the red end of the compass back to north, working on the principle that, in magnetism, like repels like and unlike attracts unlike. He can vary the strength of the field magnet by inserting it nearer or farther from the needle, suitable slots being provided for this; or he may use a smaller magnet until the error due to deviation is reduced to one degree or two degrees. He repeats this process with the machine

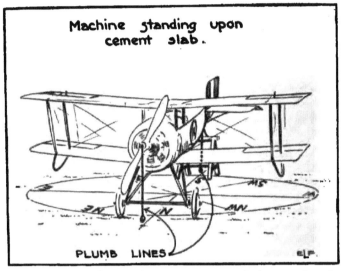

How to arrange the machine for compass swinging; testing the fore-and-aft alignment and lateral level.

heading magnetic east and west, the only difference being that, when the machine is heading east and west, the field magnet inserted must be placed in the fore-and-aft line of the machine and not athwartships.

Making a Deviation Table

Having reduced the error on the cardinal points to a minimum, a deviation table is prepared giving the compass reading for the cardinal and quadrantal points, and also stating the amount of error in degrees east or west at each of these points. In calculating his compass course, the pilot must allow for this error, and if his course lies between any two of these points he can divide the error between them. For example, if the deviation is 2 degrees east on north and 2 degrees west on east, if his course is 45 degrees his deviation will be nil. A compass having been swung for

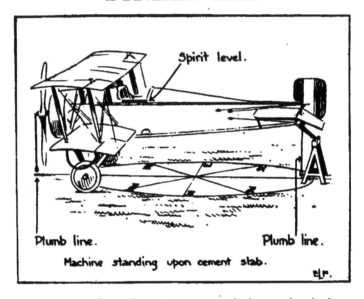

Spirit level.

Plumb line.   Plumb line.

Machine standing upon cement slab.

How to arrange the machine for compass swinging; testing the fore-
and-aft level of the machine.

deviation, no metal should be added to the machine; in
other words, the pilot should not place a lot of tools in
his pocket, or place a bag of tools behind the compass.

A table to be pinned up in the machine would look something like this:—

For Magnetic Course.		Steer by Compass.		Deviation.		
N.	0 degrees	357 degrees		3	degrees	E.
N.E.	45 ”	47 ”		2	”	W.
E.	90 ”	90 ”			Nil	
S.E.	135 ”	137 ”		2	degrees	W.
S.	180 ”	183 ”		3	”	W.
S.W.	225 ”	223 ”		2	”	E.
W.	270 ”	270 ”			Nil	
N.W.	315 ”	317 ”		2	degrees	W.

The Wind Factor

The only other factor that has to be taken into account is the wind. From his navigation lectures a pupil can make an estimate of the speed of the wind and its variation in strength and direction at the height at which he intends to fly. He can incorporate this in the course he is working out, and, assuming that he knows the speed of his machine, he can also work out the approximate time the journey will take. Supposing that the pilot wishes to fly from B to A.

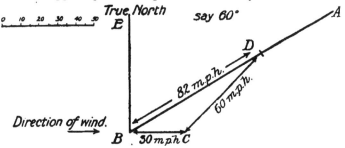

Calculating the effect of wind on the course to be steered across country.

He draws a picture and makes a scale of miles per hour, the line B—A being laid off at the correct angle to the true north for his true course. Imagine a west wind blowing at 30 miles per hour. He draws a line from B in the direction that the wind is blowing, i.e., from W. to E., and marks off on that line a point C, distant the equivalent to 30 miles per hour on his scale. Next, with centre C and radius equal to the air speed of the machine, i.e., 60 miles per hour, obtained from the scale, he describes an arc to cut the line B—A at a point, say, D, joins C—D. which will be the direction to steer, allowing for a west wind of 30 m.p.h. If he measures the angle that C—D makes, with the true meridian B—E, he will find that it gives him the correct course to steer allowing for the wind; in this case it is 44° in place of the true course of 60° obtained without reference to the wind. By measuring off B—D from his scale he can obtain his speed over the ground, allowing for the wind. In this case it is 82 m.p.h., as against 60 m.p.h. in calm air. In calculating the speed and direction of the wind, which is always given as "True," the pupil must remember that the velocity of the wind increases with height, and that at 3000 ft. it may be twice as strong as on the ground, and may have veered as much as 20 degrees.

The speed of the wind and its direction are best obtained from the nearest meteorological office, but what is known as the Beaufort Scale is often used for determining the speed of the wind near the ground. It is given on the next page for reference.

In a long flight of several hundred miles a pilot cannot have too many checks upon his journey. He must also remember the time the machine can remain in the air, and for this reason it is wise for him to get into the habit of making these calculations even when proceeding on short cross-country journeys.

Radius of Action

This presupposes that he is already familiar with the quantity of oil and petrol carried by the machine and the hourly consumption of fuel. In out-and-home flights he should be able to work out his radius of action, not forgetting to allow for the wind and the time taken in attaining his height, otherwise he may run out of petrol on the return journey.

For instance, the pilot may require to find the radius of action of his machine. If there is no wind the calculation is simple, assuming that the speed of his machine is 60 m.p.h. and it can remain in the air for six hours. His radius of action will be :—

$$\frac{60 \times 6}{2} = 180 \text{ miles.}$$

If there is a wind, however, an allowance must be made for it. Assume that the wind is dead against him on the outward journey and with him on the return. Speed of wind 25 m.p.h., speed of machine 60 m.p.h., duration of machine six hours.

THE BEAUFORT SCALE.

A System of Calculating the Speed of the Wind.

Force.	Description.	Characteristics.	M.P.H.
0.	Calm	Smoke rises vertically	0
1.	Light air	Direction of wind shown by smoke drift, but not by vanes	2
2.	Slight breeze	Wind felt on face. Leaves rustle; ordinary vanes moved by wind	5
3.	Gentle breeze	Leaves and small twigs in constant motion. Wind extends light flag	10
4.	Moderate breeze	Raises dust and loose paper. Moves small branches	15
5.	Fresh breeze	Small trees in leaf begin to sway. Wavelets form on inland waters	21
6.	Strong breeze	Large branches in motion, whistling heard in telegraph wires	27
7.	High wind	Whole trees in motion; inconvenience in walking against wind	35
8.	Gale	Breaks twigs of trees and generally impedes progress	42
9.	Strong gale	Slight structural damage occurs, chimney pots and slates removed	50
10.	Whole gale	Seldom experienced inland; trees uprooted; considerable structural damage	59
11.	Storm	Very rarely experienced, accompanied by widespread damage	68
12.	Hurricane		Above ·75

On the outward journey his ground speed will be 60 — 25 = 35 m.p.h.

On the return journey his ground speed will be 60 + 25 = 85 m.p.h.

The ratio of the speed of his outward to his return journey will be as 35 is to 85 or as 7 is to 17. Therefore the time he must take on his outward journey is 17-24 of six hours, and on his return journey 7-24 of six hours. Therefore the outward journey will take 4¼ hours and the return journey 1¾ hours. The speed of his machine on the outward journey is 35 m.p.h., which, multiplied by the time of 4¼ hours, equals a radius of action of 148¾ miles. The speed of the machine on the return journey is 85 m.p.h., which, multiplied by the time 1¾ hours, equals a radius of action of 148¾ miles.

The wind will hardly ever blow directly along the desired course, so that to work out the radius of action for oblique winds a picture must be drawn as was done in the case of allowing for drift. The ground speed on the outward and return journeys can be calculated by the formula :—

$$\text{Radius of action} = \text{petrol hours} \times \frac{\text{speed out} \times \text{speed in}}{\text{speed out} + \text{speed in}}$$

For instance, the pilot requires to find his radius of action to the S.W. on a 60 m.p.h. machine, air duration six hours, and with an east wind blowing at 24 m.p.h. As before, he draws a scale of m.p.h. (See sketch.) He takes the true bearing 225° from A and draws a line on it towards B. He produces this line in the opposite direction, i.e., N.E. He desires to fly as far in the direction of B as possible, at the same time leaving enough petrol to return to his starting point at A. Next, he marks off the direction of the wind from A, i.e., to the W., by means of a line on which he marks its strength by referring to his scale. This will be the point C. With centre C and radius (obtained from the scale) of 60 m.p.h., the speed of the machine, he cuts the line B—D both above and below A. These points are found at B^1 and D^1. Now the angle that CB^1 makes with the true north line will be the course outward, which equals 212°, and the angle that CD^1 makes with the true north line will be the course homeward, which equals 58°. The distance AB¹ equals the outward speed, e.g. (obtained from scale), 73 m.p.h., and the distance AD¹ equals home-

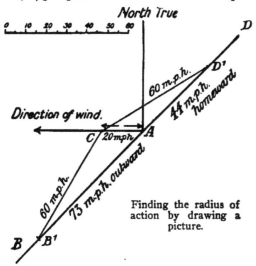

Finding the radius of action by drawing a picture.

ward speed, e.g. (obtained from scale), 44 m.p.h. Then, by applying the formula already given, he works out his radius of action as follows:—

$$\text{Radius of action } \frac{6 \times 73 \times 44}{73 + 44} = 6 \times 73 \times 44 \quad \frac{}{117} = 164.7$$

miles. This result allows of no margin of safety for a change of wind to a more adverse quarter or for time lost should the pilot lose his way, so that in his calculations a margin of at least 20 per cent. should be given.

All these calculations will require to be verified in the air for safety sake, but the pilot will be surprised how accurately it is possible for him to work out his speed and course.

Finding One's Bearings

Having worked out the course, it will be found useful to divide the line on the map which joins A and B into equal distances, each representing, say, 10 miles on the ground; then, assuming that the pilot has calculated that his speed over the ground will be 60 m.p.h., if he takes his time of departure, even if he gets lost, he can tell approximately where he should be on his journey by looking at his watch. This will help him to find his bearings again, as a lot of ground is eliminated from his reckoning. Suppose, for instance, that a pupil lost himself for some reason after flying for about 30 minutes. He will know, if he has divided up and numbered his line, that he must be somewhere about the third mark. All he has to do then is to find some landmark in that neighbourhood and recognise it on the map, or vice versa. If the pupil realises that he has not sighted one or two principal landmarks that he should have done, he should come down and ascertain where he is. Of course it is of the utmost importance that all these preliminary calculations should be accurate, and if he calculates his course as north-east when it should be south-west, or reckons the wind is with him when it is against him, he is sure to make a mess of things; hence the importance of checking preliminary calculations very carefully, as these mistakes are more common than one would imagine.

The Inter-relationship of Landmarks

He will now spend some time in studying his map, and, if possible, he should be able to memorize the principal landmarks, the order in which they will present themselves to him, and their relationship to one another. He should note the highest ground he has to fly over and should know the kind of landmark to look for. In cross-country flying he must remember that hills and valleys look flat at a height of a few thousand feet; therefore, he should not waste time in trying to memorise them. It is such features as railways, large sheets of water, reservoirs, lakes, rivers, woods, towns and sea coast that he should pick out. More important still is the inter-relationship between these features that he must visualise. He must study how railways and rivers enter and leave towns on or near his route; how railways join each other or intersect; any curious curves or particularly straight stretches of railway line, river or canal; the angles that towns, sheets of water or woods bear to each other; and, by doing this (which will come natural to him after a few cross-country flights), he will become familiar with the details of the journey ahead of him even before he takes the air.

The Chief Characteristics

It is a good plan, too, on a long flight to study the kind of country to be crossed. Thus the first portion may be suburbs, and then may come 10 miles or 20 miles down the valley of a big river, where the ground is flat and marshy. The next stage may consist of thickly-wooded country, and the last stage, perhaps, of rolling downs and open farmland. If he lost himself on his journey, he could probably tell his whereabouts by the nature of the land under him. The importance of this preliminary study of the map by a pupil cannot be sufficiently impressed on him, and, to start with, perhaps, he will do well not to fly too high until he is proficient in map reading and realises, for example, that a town which, at 1000 ft., he can recognise street by street and house by house, looks like a black smudge when he ascends to 8000 ft. or 10,000 ft.

Preparing and Studying the Map Before a Cross-country
Flight

An Explanation of the Accompanying Map

EXAMPLE.—Flight from aerodrome B, near Cardiff, to
aerodrome A, near Aberystwyth.

Join A B, which will represent the line of flight.

From the scale at the foot of the map find the length of
the journey = 74 miles.

Divide the journey into equal sections of, say, 10 miles;
mark and number these on the map.

Work out the compass course by drawing line B C true
north (obtained from true meridian) and measuring angle
CBA = 329 degrees. Apply variation, say 15 degrees west,
in which case it is added = $329° + 15° = 344°$. Also,
apply deviation from deviation card in machine, say 5 degrees, east, in which case you subtract. Compass course
then equals 339 degrees. A figure can be made out to calculate for wind as described elsewhere, but in this case it
may be assumed that there is no wind.

Assuming the air speed of the machine to be 60 m.p.h.,
the pilot calculates that the journey should take about 1¼
hours, and that each stage of 10 miles will be covered in
10 minutes. Supposing he starts at 12 o'clock midday, he
will pass the stage marked 1, at 12.10; stage 2, at 12.20;
stage 3, at 12:30 and so on.

Cut the map with scissors or a knife the correct width to
fit in strip form in a roll-type map-carrier, and mark the
course and distance on it. [The section of the map printed
is so cut.] Then try and visualise the map and note the
important landmarks, their relation to each other, and at
what time and distance they should come into view. Note
also the fact that most of the ground traveled over is mountainous, and therefore bad for landings. On stages 1, 2, 3
and 4 there are mountains 1500 to 2000 ft. high, but the
worst ground, judging from the map, occurs between the
stages 5 and 6. The very unfavourable character of the

139

ground necessitates the pilot flying as high as possible, so
that he will probably add 10 minutes or so to the time of his
first stage to allow him to gain a good altitude of, say, 5000
or 6000 ft.

Study the main features and landmarks of the map stage
by stage.

Stage 1.—Caerphilly railways in X form make an excellent landmark; also the big wood S E; note the shape of
this wood and also the water at Llanedeyrn and Whit-
church.

Stage 2.—Note the double railway lines, road and water
leading direct on the course to Merthyr Tydfil, which should
be easy to pick up, but which may be confused with Rhymney or Aberdare if the pilot is off his course at all. The lake
to the W. of Aberdare would show him his mistake, however, but here again it may be confused with the lake W.
of Rhymney.

Stage 3.—The road and two sheets of water on it are the
best guides in this case. The pilot might note the strag-
gling wood to his left and several sheets of water to his
right, also the railway curving away to the N.E. from
Merthyr Tydfil.

Stage 4.—If the pilot has got so far he cannot miss crossing the Meath and Brecon railway, and he should try and
make for the point where it joins the River Usk. The
wood Yr Allt is an additional landmark at this point, and
he should note its shape and relation to the river and railway. The large reservoir on his left is another good landmark.

Stage 5.—Note Llandovery, the river and woods on the
left, and fly on until crossing the railway, taking note and
not being misled by the tunnel. If in doubt then keep along
on the compass course until crossing the banks of the River
Towy, where there are woods on each side of it. Note
Llanwrtyd Wells, the road, railway and river on right, also
the wood N.W. of the town.

Stage 6.—Nothing much to indicate the route except one
small lake on the right and possibly woods round Lampeter,
if the visibility is good. A difficult section for this reason
and also because of the mountainous country traversed.

Stage 7.—Tregaron forms an excellent landmark. Note

the V fork of the roads, also the railway, which must be
crossed at right angles. Then keep to the left of all woods
and railways. Look out for the sea coast on your left and
fly along it northwards until striking the first large town. If
the pilot strikes the coast too high up he can easily find
his mistake out as there is a railway running parallel with
the coast to Aberdovey, whereas S. of Aberystwyth there
is no railway. He should not mistake Aberdovey for
Aberystwyth, because the mouth of the Dovey is a distinctive feature in the former town and the railways are
quite differently arranged. Having studied the map as indicated, he can fit it in his map case, roll it up and get ready
to start.

To assist the pilot, a rough time and distance table of the
journey may be made as follows :—

Time.	Distance. Miles.	Landmarks.
12 o'clock mid-day	—	Start. Climbing for height.
12.10	7	Caerphilly Wood S X railways.
12.30	20	Merthyr Tydfil, cross-roads and double railways and road direct on course.
12.40	30	Two lakes on course.
12.47	37	Cross railway, roads and river.
12.54	44	Llandovery, river and woods on left.
1.08	58	Round lake on right.
1.10	60	Tregaron, railway and roads crossed at right angles. Sea coast should be in sight very soon.
1.24	74 miles	Arrive Aberystwyth, dense woods on right.

To make certain that the table is correct the pilot could
time himself over one of the stages and work out his speed
from that. The woods, coloured green in the original, are
difficult to reproduce, and are, therefore, not easily discern-
ible in the accompanying map, and allowance must be made
for this.

How to Carry the Map

The next thing to do is to fix the map in the most convenient position.   Some pilots prefer to have it pinned up in front of them on the instrument board, and others to have it slung round their necks in a roll-type map carrier.   In either case the line of vision of the pilot should be removed as little as possible from straight ahead when he requires to read his map.   This remark applies equally to instruments, which are better fitted in a group than when spread out over the dashboard.   A third method is to carry the map loosely in the pocket; but it may easily be lost, and, in any case, it is difficult to unfold or turn over when necessary.   If the map is carried in a roll case slung round the neck, care should be taken that the case does not jam the control lever at any time.

The pupil is now almost ready to set forth on his flight. Where it is possible, and the weather is changeable, it is a good thing to telephone to aerodromes en route, and also to his destination, and obtain a weather report, which will give such items as height of clouds, strength and direction of wind, and visibility.   He must glance at the barometer, too, and if the glass has fallen much he must look out for bad weather.   Incidentally, the height indicator can be used as a barometer, and if set to zero one evening and found to read 400 ft. the next morning this indicates that the glass has fallen.

Preliminary Details

While waiting for the weather report, the pilot can look over his machine and engine as described previously.   He must see that the petrol and oil tanks are full and that the instruments are working properly, that the height indicator is set to zero, the engine running well, and a bag of tools with a few spare parts, such as sparking plugs, valves, magneto contact breaker, some high-tension wire, a complete petrol pipe with unions, some insulating rubber tape and copper wire, are securely packed away in the locker.   None of these things should be taken for granted, and it is no insult to the mechanics to look over the machine one is going to fly.

In some cases in England it is necessary to send warning to the air authorities and to obtain leave from them before starting on a flight. It is advisable to find out whether the flight necessitates crossing any defended areas, which can be found out by telephoning to headquarters. These should generally be avoided, unless they have been advised of the flight before, otherwise the pilot may be mistaken for an enemy aviator and dealt with accordingly.

If the pilot is in the Service he must remember to take his Service cap with him. This can be stored in one of the drawers or lockers fitted to most machines. If a suit case is carried, care must be taken to have it firmly secured to the seat so that it cannot possibly foul the engine or machine controls.

Position and Comfort

It should be remembered that a cross-country flight of an hour or more is a much more tiring undertaking for a

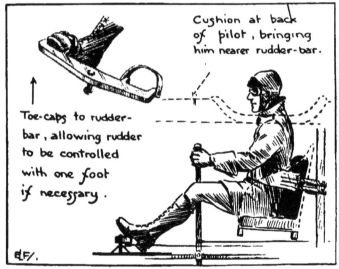

Cushion at back of pilot, bringing him nearer rudder-bar.

Toe-caps to rudder-bar, allowing rudder to be controlled with one foot if necessary.

A comfortable position for the pilot is most important in long flights. Two points worth remembering are illustrated.

beginner than a few short trips around the aerodrome, although for his instructor the reverse is the case, as a straightforward flight is much easier than continual landings and turnings. The pupil should make himself as comfortable as possible in his machine. He should avoid cramped positions at all times, and, if necessary, the seat can often be moved further back to give him more leg room. In view of the possibility of cramp in the legs, it is a good plan to fit toe straps on the rudder bar, so that the rudder can be controlled by one foot while the pilot stretches the other. A short man should make certain that he can reach the rudder when it is fully over with his knee just bent, and if he cannot he should fix a cushion, or cushions, behind his back so as to pad him up. Similarly, if his view is obstructed, he can improve it by adding an extra cushion or two to his seat.

Clothing

Plenty of clothing should be worn even in summertime. Leather overalls are the best for general work, and sometimes it is possible for a pupil to draw a flying kit on loan. This can be supplemented in winter-time by woollen sweaters, mufflers and helmets. The hands and feet are the parts of the body most likely to be affected by cold. Big boots are essential, and with a combination of silk and woollen socks or stockings should help to keep the pilot warm. Goloshes or snow boots are also excellent, although somewhat clumsy. Many pilots object to the wearing of cumbersome gloves or boots on the ground that it interferes with the touch and feel of a sensitive machine; but in the event of their hands or feet becoming frost-bitten through insufficient protection, it may be remarked that they will have no feel at all. Fur gloves lined with wool are as warm as anything, and a fur flying cap, fitting close over the head, without ear holes, provides a suitable headgear and one on which it is possible to raise and lower the goggles easily. Sometimes a pilot takes two pairs of goggles with him on a very long flight in case one pair should get lost, for it is a very bad thing to fly without goggles at any time. This is especially so on tractor machines lubricated with castor oil, which is most injurious to the eyes. The goggles should fit well and admit no draughts. Those fitted with Triplex glass are the best because, in the event of a smash, there is less chance of the glass breaking and cutting the pilot's face or eyes. It is a good plan to carry a rag to clean off the oil and dirt that collects on the goggles and windscreen from time to time, or the pilot may use the loose end of his muffler for this purpose. Newspapers provide an excellent protection against cold, and may be worn under the waistcoat or in trousers or boots. It goes without saying that damp footgear should be avoided at all costs, more particularly by those who are prone to chilblains and bad circulation. They may receive some comfort by greasing their hands and feet before starting.

One last word before the pupil takes the air. In view of possible forced landings, it is just as well to take plenty of money on the journey. This will not only make the period of waiting for relief more pleasant, but will also facilitate repairs if made on the spot with the aid of local motor men or carpenters, as well as the possible guarding of a machine by a farm labourer.

The pupil should know the telephone number both of the place he has left and of the place to which he is going, in order to ring up either aerodrome when he requires assistance. A pilot's ticket and photograph issued by the Royal Aero Club forms a useful means of proving his identity. Some form of identification should always be carried.

Fitness for Flying

A pupil should never proceed on a cross-country flight if he does not feel physically fit, and he should have no hesitation in telling his instructor if he does not feel well. This also applies when the pupil is to make his first solo, as then he wants all his faculties and muscles working harmoniously, quickly and accurately together. There is no question of cowardice in a pupil saying that he does not feel fit or well enough to make a flight when in the early stages of tuition. The instructor will be able to judge from his previous experience of the pupil whether the man is really unwell or only pretending.

CHAPTER IX

Cross-country Flying

THE pilot is now ready to take the air on a journey from B to A. Before starting, he notices the time of the day, or sets his watch to 12 o'clock, or to the nearest hour, so as to facilitate future reckonings.

It is a good plan to climb to 1500 ft. or 2000 ft. before leaving the aerodrome, and then to set off the wind allowance, which can be compared with the course already found by calculation. This is done by taking some point on the line of flight already determined and one which is visible from the aerodrome. The pilot then flies on this line of sight and notes the reading of his compass, which will give him the course to be steered. This is a very simple and, at the same time, effective method of finding out the compass course without any mathematics or reckoning whatever. As an alternative, he can fly on his compass course, which includes his allowance for wind, and then note if it takes him along the line of country it should do on his map. If it does not, he must alter his course and note the bearing.

The Height to Fly

Generally, when operating over friendly territory, any height between 2000 ft. and 5000 ft. is suitable, depending on the conditions of clouds, visibility and landing grounds and the highest portion of ground to be crossed. For instance, if a range of hills 1500 ft. high had to be crossed, and a start was made from sea-level, the pilot would not fly under 3000 ft. in their vicinity. An eye must be kept open continually for landing grounds, while the direction of the wind must be noted. Assuming that the machine has a gliding angle of 1 in 7, in still air a pilot can select any landing ground within 3000 yds., or, roughly, something under 2 miles, from a height of 1285 ft. Therefore, the suitability of the ground that is to be flown over must, to some extent, regulate the height attained. In flying across the Channel or a very large city, a height of 10,000 ft. or more would be reached.

Having determined this question, the pilot then climbs slowly to the height at which he decides to fly. Avoiding clouds, either by flying under, over or round them, the pilot should, at any rate in his early flights, keep the ground in view the whole time, steering by compass and picking

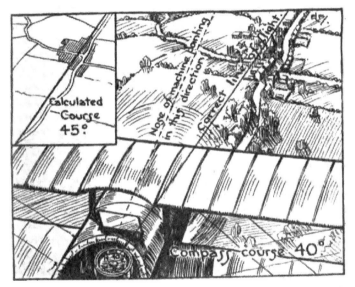

When the pilot has worked out his compass course, he can verify the accuracy of his calculation by flying by sight to a landmark that he knows is on his correct course, and checking the reading of the compass with his calculated course. In the illustration the pilot calculated his course to be 45 degrees, as shown in the small map inset. When he got into the air he found that his compass read 40 degrees when flying by sight on the correct line of country, i. e., to the first village. This indicates a gentle northerly wind, for which he had not allowed. The line of flight made good is indicated by the thick, black line.

Flying in clouds. The pilot of the top machine is maintaining his lateral level by the cloud bank in front of him, although the ground horizon is obscured. The lower machine is descending through a hole in the clouds, through which the ground is visible.

up landmark after landmark as each presents itself to his vision. The higher he goes at first the more difficult will it be to recognise the ground under him by the map.

Storms, Fog and Clouds

There are many difficulties of various descriptions with which he may be faced. He may run into a rain, sleet, or hailstorm. These will generally be passed through, but hail may chip his propeller slightly. If a snowstorm is encountered, the best thing to do is to turn back or land at once. Fogs or mists should be avoided and a landing made at once when there is some danger of a pilot becoming enveloped in one. Heavy banks of clouds can generally be avoided, and for pupils this is the best course, for even the cleverest pilots are liable to lose their bearings in such circumstances. If a pilot should lose sight of the ground, and yet have clear air above, the machine can be maintained on an even keel by using as a horizontal level the very banks of clouds which hide the ground. Generally, these banks are more or less level, and, even if they are not, the pupil can take some well-defined cloud formation and maintain his position relative to it. If it is necessary for a pilot to fly through clouds or fog, he must watch his instruments most carefully and fly by them alone as best he can. He should brace his rudder square and try to keep his control lever central and vertical. If he is above the clouds and wishes to descend, he must first of all look round for any holes through which he can see the ground. In gliding down through these holes he must be careful to look out for other school machines which may be flying under the clouds. If there are no clear places in the carpet of clouds, he must select a spot where it seems to be thinnest and glide down through it quite steadily.

A Natural Gliding Angle

Most machines will take up their natural gliding angle with the engine switched off, and only the slightest pressure on the stick is necessary to maintain them in a steady descent. A pupil will find that being in a cloud is very much like being in a really thick mist on the ground. Everything is damp and dark. In fact, in some clouds it is so dark as almost to prevent him reading his instruments. His aim should be to keep on his steady gentle glide until he sees the earth again. He should keep a careful watch on his altimeter, and if he gets to within a few hundred feet of the ground and there is no sign of the cloud thinning, he can reckon that he has been caught in a mist or fog, and had better climb again, get up over it and look for its limits, which may be in a down-wind direction. He must fly down wind by compass and try and overtake the advancing fog. He must also keep an eye on his petrol supply, for if he were forced to land in a fog through lack of petrol his plight would be even worse than if he had to land with the use of the engine, which might save him from running into houses or trees, seen at the last moment.   Ground mists and fogs are undoubtedly the most dangerous and treacherous enemies of the aviator.   There is no particular difficulty in climbing through banks of clouds.   The pilot can twist and turn his machine upwards through the clouds, always making for the brightest spot, until, sooner or later, he will emerge into clear blue sky and perfectly calm air. Quite often, really bad bumps are experienced in and near clouds, and it is these that may upset the pupil's idea of balance in climbing or descending, so that he may emerge from them in some unusual position, which will not matter in the slightest if he has sufficient height to right the machine again.   Many pilots have noticed how their compass card appears to spin in a cloud, and they have attributed this peculiarity to electrical effects, although it seems more likely that it is their machine which is spinning, whilst their compass card is trying to point to the magnetic north during these evolutions.

When Clouds Obscure Landmarks

It is much more difficult to study the ground when it is cloudy than on a clear day, because the pilot can only see little bits of road, railways or woods, and it may happen that the very landmark for which he is searching to give him his whereabouts is obscured from him.   For the same reason, it is very much more difficult to fly across country and find the way when visibility is poor than when it is good and permits high flying.   The pupil might think that the lower he flies the easier it is to recognise the ground; but the opposite is the case, because he does not want to recognise the ground house by house, or mile by mile; he wants to be able to see long stretches of road, river or railway, and to see how they curve and intersect at distances of 10 miles or 20 miles from his position.   The further he can see ahead the easier he will find his map reading, and this means that he must have good visibility.   After all, the map is a picture of the ground from a great height.   He can see all his map at once, but he cannot, unless he be very high and the visibility be excellent, see all the ground that it depicts. If he could, there would be no need to use a map at all. He would get up high over his starting point and would fly by sight to his destination, even though it was 50 miles or 60 miles away.

Atmospheric Conditions

Another reason for flying fairly high is that the air will probably be much calmer there, and probably he will not be worried by heat or wind bumps. Early morning or late evening are the best times for flying, for the atmosphere is both cooler and calmer then. Heat bumps generally begin to manifest themselves soon after sunrise, especially in summer. To begin with, they may be within 100 ft. of the ground. They gradually ascend, until at mid-day they may extend to 2000 ft. or 3000 ft. high. Bumps are generally found in the vicinity of clouds and inland sheets of water, although at sea the air is usually calm. In fact, a wind from the sea is often a calm wind, in the sense that it is not bumpy, although it may be blowing strongly. Pupils should get used to flying in bumps as soon as possible, as the more they do it the less uncomfortable will they feel. In landing in a really strong and bumpy wind, it is a good plan to come down with a little extra speed, as they will have more control of their machines; but it must be remembered that, owing to the wind, the ground will appear to be coming towards the machine during the flattening-out process, very much more slowly than when landing in a calm wind, and the pupil must make allowance for this. Generally a pupil finds more difficulty in getting off in a strong wind than in gliding and landing. In landing in a strong wind in the vicinity of hangars, buildings or woods, the pilot must look out for sudden gusts and eddies set up by the wind striking these obstacles. These gusts may cause the machine to roll or drop 10 ft. or 15 ft. Hence the need for giving as wide a berth as possible to buildings whilst gliding down to land.

It is very unlikely that a pupil will be sent off on his first cross-country journey under unpromising weather conditions. In all probability he will have nothing much to do except to steer his compass course, read his map and come down when he decides, landing in the aerodrome A just as he would at his own at B.

Taking Bearings

Suppose that a pilot loses his bearings. There are a number of ways in which he can find his whereabouts. He should know where he is approximately by calculating his speed and the time he has been flying, and then searching the country indicated on his map at the approximate point on his course which he should have reached. If his whereabouts are not to be found by this method, he must then circle round some clearly-defined spot on the ground and try to recognise it on the map somewhere near where he ought to be at that time, bearing in mind that, as explained in the chapter on map reading, it is the inter-relation of landmarks and main features that is the easiest to recognise. Another method is to come down low and read the name of a railway station, which can be done from a height of 500 ft. to 1000 ft., depending on the light and lettering on the nameboard.

As a final resource, he may land and inquire the way. It may happen, however, that, although he lands in friendly country, he is unable to find out where he is from the inhabitants; either he may not understand the language, or else there may be nobody there to help him. He must then study his map, proceeding as follows:—He must first set off from a side line on the map an angle representing the variation of the compass (in England this is generally 15 degrees west). This line will be the magnetic north line. He must then twist the map round until the line that he has just drawn lies on the magnetic N. and S. line, as shown by the compass in the machine. This is called setting the map. He must compare the surrounding country with the map and try to recognise the landmarks. It may be the intersection of a road and river, or cross railway lines, or a town, church, range of hills or woods. He must recognise these places on the map, and, from their bearings to each other, work out his own position. Sometimes he

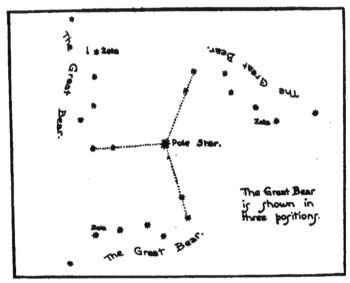

The position of the Great Bear constellation in relation to the North Pole Star.

can find this by walking to the nearest road and reading a milestone or signpost.

If he is unable to use his compass to assist him, he must recognise two points on the ground and on the map, and then place himself between them. If he joins the line between these places on the map he will be somewhere on that line. By taking two more places in a similar manner, so that a line joining them on the map will be as nearly as possible at right angles to the first line he drew, he will discover his position at the intersection of the lines on the map.

To find the true north, the watch method already described is useful, whilst at night the north star is the best guide. This can be found by locating the seven stars forming what is called the "Great Bear," the "Plough," or the "Saucepan." The two end stars at right angles to the others are called the pointers, and the north star will be

found in line with them, but at a distance roughly five times
the space between the pointers.   In England it is worth
remembering that the aisle of most churches is built east
and west, the altar being set at the east end.   The sun
also will help a pilot to find his position, as it rises in the
east, is at the south at mid-day, and sets in the west,
these pointers, of course, being only approximate.

Landing on Strange Ground

If it is necessary for a pilot to land off an aerodrome
in order to find his way, or for any other cause, but not
because his engine has failed, he will make more certain
of hitting off his field by·flying into it and using a little
engine.   As soon as he sees that he is going to land in the
field safely he can throttle off.   To make certain that the
field is suitable, he should fly low once over it, scrutinising
it for hidden pitfalls.

Obviously, when landing on strange ground he should
choose a field free from obstacles, for if he has to come
in over trees or houses it will mean that there will be much
less room for him to land.   Here the best method of bringing down a rotary-engine machine may be described.   The
petrol tap should be shut, but the adjustable needle valve
which regulates the mixture, and also the switch should be
left alone.   The sparking plugs will not soot up so readily
when the switch is left on, and then, when the pilot desires
to open out his engine again, he has only to turn on the
petrol and his mixture is automatically correct, unless he
has come down from a very great height, in which case a
slight adjustment to the needle valve will compensate for
the change in the density of the atmosphere.   The rule
is that the higher the machine is the more must the petrol
be cut down, so that, on descending, he will have to give
a little more petrol to correct the mixture.   He should
open out his engine at a height of about 1000 ft.; and
can then switch on and off until he is certain that he is
going to make the field.   Then, especially on a fast machine, he should leave the switch off altogether while com-
pleting the glide and landing.

In bringing down a machine fitted with the Monosoupape engine, the switch is left on and the fine adjustment closed. The machine can be landed with the engine off and the fine adjustment is opened up to half normal running position for taxying. The switch can be used if desired. In "blipping" an engine on the ground, a pupil should practise switching on as little as possible and allowing a long interval between the "blips." He should use as little petrol as possible, as it is obvious that, on a rotary engine (the only type where blipping is necessary), much less petrol is required to keep the engine running at intervals compared with that required to keep it running continually. Hence the fine adjustment must be half closed when the engine is being controlled on the switch. Even in the air, the object should be to use as little petrol as possible consistent with even running, and if the machine misfires or chokes, even occasionally, the fine adjustment lever must be altered to obtain perfect results.

Propeller Torque

Sometimes during the last few hundred feet the pupil may use his engine the better to gauge his distance. In switching on and off, or, as it is commonly called, "blipping" an engine, either on the ground or in the air, the pilot will notice that the lateral balance of the machine is temporarily upset and that one wing will rise and fall as the engine is switched on and off. This is due to what is called the torque of the propeller. If the propeller revolves in a clockwise direction, judged from the pilot's seat, it creates an opposite turning movement of the machine, which will tend to revolve about its longitudinal axis in an anti-clockwise direction. Unless due allowance were made for this the machine would fly left wing down. But to compensate for the torque of the propeller more lift is given the left wing by increasing the incidence, or giving it what is called a "wash in." The greater lift allows the machine to fly laterally level, despite the propeller torque. When the engine is switched off for descending, the torque decreases, and so the left wing with the greater lift will tend to rise momentarily. When an engine is being "blipped" on the ground, the same sideways rock of the wings will be noticed. When the engine is on, the wing will drop, to rise again level as soon as the switch is off.

Badly Trued-up Machines

There is another point which the pupil must study whilst flying various types of machines. Sooner or later he will come across a machine which may not be perfectly trued up, either in the fore-and-aft line or else laterally. He may have heard of such-and-such a machine flying right wing heavy or right wing low. This means that the con-

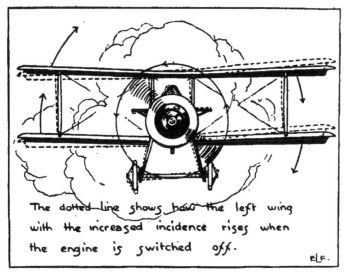

The dotted line shows how the left wing with the increased incidence rises when the engine is switched off.

The torque of a clockwise-running engine or propeller (viewed from the pilot's seat) tends to make the machine turn in an anticlockwise direction, i. e., to fly left wing down. This tendency is counteracted by a wash in or increase of incidence on the left wing, as shown by the blacked portion in the sketch. The dotted line shows what happens when the engine is switched off in the air, and causes the left wing to rise momentarily owing to the disappearance of the propeller torque.

trol lever has to be held over slightly to the left in order to make the machine fly horizontally or laterally level. If the control lever were placed central for flying, as it ought to be in the properly trued-up machine (which should almost fly and glide itself, hands off), the machine would travel along right wing low. The cure for the trouble is either to increase the incidence on the right wing or to adjust the aileron controls. The same trouble might equally apply to the left wing.

Setting the Tail Plane

A somewhat similar complaint results in the machine flying nose or tail heavy. If, in order to fly level in the fore-and-aft direction, the pupil has to hold the control

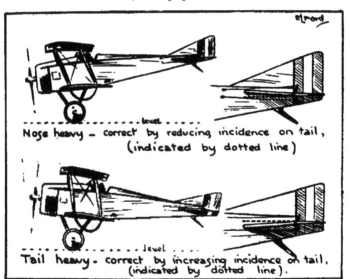

Examples of nose-heavy and tail-heavy machines, showing how to alter them by adjusting the incidence of the fixed tail plane. On some machines it is possible to vary the incidence of the tail plane in the air, and thus trim the fore-and-aft balance of the machine from the pilot's seat.

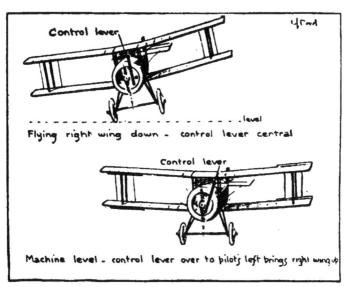

Examples of machines flying right wing low or right wing heavy.

lever forward with a certain amount of pressure, the machine is said to be tail heavy. The cure for this is to give more incidence to the fixed tail plane, which will correct the fore-and-aft balance of the machine. A nose-heavy machine has to be held up by the control lever in order to fly level, and is remedied by the tail plane being rigged with less incidence.

Restarting from a Field

When restarting, the longest possible run should be taken, and if there are trees or houses at the far end it might be wiser to get off across wind rather than to chance getting off over the houses. It is quite easy to get off across wind, although it is a proceeding that pupils should not attempt until they are more or less confident in their flying capabilities. Supposing a machine is getting off side to wind, with the wind blowing from left to right, the pilot must keep the rudder over to the right, judging the amount according to the strength of the wind, and, fixing his eye on some point ahead of him, must steer his machine towards it, using right or neutral rudder and left bank, so as to attain this object. Once the machine is off the ground there is no particular difficulty in the manœuvre, and the pupil can turn into the wind as soon as he has gained a little height. It would be inadvisable to turn down wind until quite a considerable altitude, say, 1000 ft., has been attained, owing to the difficulty of bringing off a forced landing if the engine failed. In landing across wind, the same procedure is followed, the wing nearer the direction of the wind being canted towards it slightly, which almost has the effect of making the machine land on one wheel. In practising this, the pupil will notice that what really happens is that the drift, or crabwise motion, of the machine over the ground, due to the side wind, is compensated for

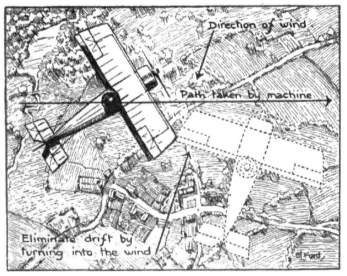

The course of the machine over the ground is often crabwise, owing to the effect of a side wind. When the pilot wishes to land into the wind he must eliminate the drift or crabwise motion by turning into the wind, as shown by the dotted aeroplane.

by a slight inward sideslip, which allows the machine
pursue a straight course over the ground.

If there is ever any doubt about clearing an obstacle
getting out of a field, it is as well to remember that by t.
ing off weight, such as by shedding the passenger, to
cushions, or even running off some of the fuel, the mach
will take the air quicker and a great deal of the "run" c

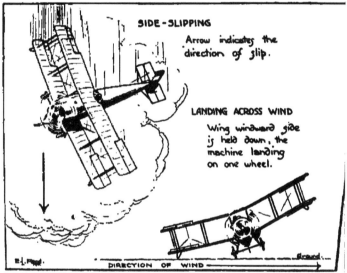

In the position shown on the left the machine will sideslip. Landing
across the wind is rendered quite easy by landing with one wing
down, and the sideslip into the wind which follows eliminates a
crabwise course.

be saved. All this may have to be done to prevent a machine having to be taken to bits and packed up as the only
means of getting it out of a field. It happens sometimes
that an instructor has to fly over to assist some pupil, and
if the pupil is in a bad or difficult field the instructor should
not attempt to land there, but must pick the nearest suitable field. If he lands in a field where there is a doubt
about getting out again, he can shed his passenger, get out

f the field solo, and pick his passenger up again by land-
ng in a more suitable field near by.

Before restarting in any strange field the pupil will do
well to make a journey on foot to survey the ground in
front of the machine. In extreme cases, crops or part
of a hedge may be cut down in order to allow a machine

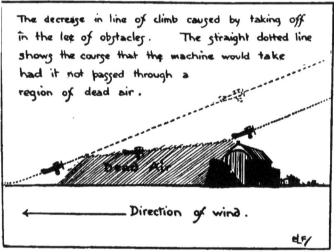

The decrease in line of climb caused by taking off
in the lee of obstacles. The straight dotted line
shows the course that the machine would take
had it not passed through a
region of dead air.

Direction of wind.

A point to be noted in taking off in the lea of obstacles. The region
of the dead air which is shielded from the wind by the shed is
shaded, and the course of the machine passing through it is indicated by the lower dotted line. The straight, upper line shows the
angle of climb attained by the machine had it not passed through
the region of dead air.

to get off safely. If a pilot is faced with a small obstacle,
such as a hedge at the end of a small field, the best way
is to taxy up to it as fast as possible, and then to pull
the control lever back suddenly, when the machine will
jump the hedge, even though it has hardly yet gained a
flying speed. Sometimes it happens that a machine has to
be got off in such a way that it travels through air which
is sheltered from the wind. The pilot must allow for
this, as the lift obtained from air in the lea of sheds or
woods, where the air is comparatively "dead," is not nearly

so great as the lift obtained if the machine could get straightaway into the wind. The same remarks apply to landing in a sheltered area where the machine will glide and run much further than might have been expected.

In landing in a field, the pilot should always aim to land at the near-side fence just grazing the hedge, so that he has the whole length of the field for his run.

In cross-country flying it happens sometimes that the engine fails and the pilot has to make a landing in strange country without its aid. This is called a "forced landing." One of the best tests of a pilot's skill is his ability to bring off a forced landing. Let it be assumed that the engine shows signs of giving out at a height of 3000 ft. A good

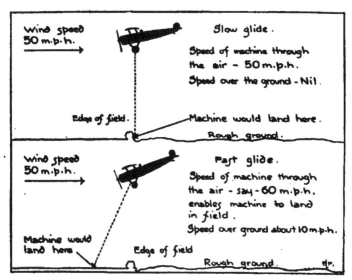

Wind speed
50 m.p.h.

Slow glide.

Speed of machine through the air – 50 m.p.h.

Speed over the ground – Nil.

Edge of field.

Machine would land here.

Rough ground.

Wind speed
50 m.p.h.

Fast glide.

Speed of machine through the air – say – 60 m.p.h. enables machine to land in field.

Speed over ground about 10 m.p.h.

Machine would land here

Edge of field

Rough ground.

Landing in a wind. Illustrating why it may be advisable to nose the machine down more at a faster gliding speed against a wind than in calm air, if it is desired to land in a certain field. In this case the air speed of the gliding machine and speed of the wind are made the same, so as to illustrate the force of the argument. It is assumed in this case that at 60 m.p.h. the machine has a more efficient gliding angle than at 50 m.p.h. At 70 m.p.h. the machine would not glide so far.

axiom to remember is that it is far better to land with some engine, even though it be but four cylinders working out of eight, rather than to land without any engine at all. The power exerted even by four cylinders may be just sufficient to prolong the glide over a hedge which otherwise would have been short, or to help a pilot to jump his machine over the next hedge if he has overshot the mark. Hence, when an engine shows signs of gradually losing its power, it is a wise plan to come down and investigate rather than trust to luck and wait for the engine to fail completely. On some engines it is by no means a rare occurrence for a partial seizure to take place. If allowed to go too far, this will cause the engine to stop dead, and it may be seriously damaged; but a landing can be made with half or three-quarter power at the pilot's control if the revolution counter be watched carefully and the sound and feel of the engine be studied. "Any motor is better than none," is worth remembering at all times. A gradual seizing-up of some engines, due to wear of the obturator ring, is indicated by the blueing-up of the four top radiating fins. This applies more particularly to the Gnome engine.

Making Use of the Wind

Most forced landings are due to total failure of the engine. As often as not this is caused by quite a small defect, such as a broken petrol pipe, faulty switch, or defective contact breaker on the magneto. Sometimes one or two plugs soot up, but they may right themselves in the course of a few minutes; hence the need for the pilot gaining some elementary knowledge of the vital parts of his machine and being able to look over these points himself before starting. When an engine fails completely at a fair height, the pilot should switch off and turn off the petrol to prevent the possibility of fire, which is more liable to occur on a rotary engine than on a stationary one. The pilot then turns his machine into the wind. He does this automatically and accurately because he has always been noting the direction of the wind during a flight with a view to possible forced landings. He has been looking for smoke from rubbish heaps, haze over towns, steam from railway engines, flags, shadows of clouds moving on the ground, and, when close to the ground, the movement of crops or the foliage of trees. If a strong wind is blowing across the machine, it is nearly always possible for a pilot to watch his drift as he glides down, always turning his machine away from the direction in which he is being blown. He must also calculate the strength of the wind, because, if there is a strong wind against him, his range of glide against it will be much shorter than if there were no wind. When a very strong wind is blowing he may have to select a field almost vertically under him, because the speed of the wind is approximately equal to the speed of the machine, which, in such circumstances, would descend vertically on its glide to earth. For this reason a pilot should have some idea as to what his most efficient gliding angle is—that is, at what air speed the machine will glide the farthest. For the sake of example, let it be assumed that a machine has a range of flying speed from 40 m.p.h. to 60 m.p.h., depending upon the amount of engine used, and that 50 m.p.h. is its most efficient gliding angle. If the pilot brings the machine down either faster or slower than this speed he will undershoot his mark. Hence the importance of his knowing how to lengthen or decrease his gliding distance.

Fast and Slow Gliding Angles

To reduce the idea to extremes, the machine would come down vertically in a nose dive at, say, 120 m.p.h., and it would also come down vertically, or almost vertically, if it were allowed to descend stalling at, say, 30 m.p.h. The net result would be approximately the same. If a pilot were gliding down slowly at, say, 45 m.p.h., with the view of decreasing his run on landing, and he saw that he was going to undershoot his field, he could reach it safely by increasing the speed to 50 m.p.h. (i. e., his longest glide), although he would be putting the nose down to do it. But if he were going to undershoot his field when already gliding at 50 m.p.h., his most efficient speed, he would undershoot it still more by putting the nose down and increasing the speed. He would also undershoot it if

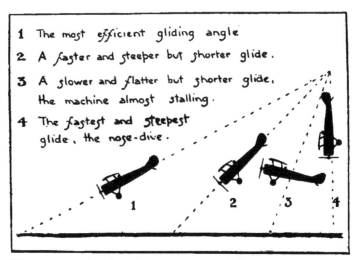

1 The most efficient gliding angle

2 A faster and steeper but shorter glide.

3 A slower and flatter but shorter glide, the machine almost stalling.

4 The fastest and steepest glide, the nose-dive.

Diagram showing that the distance travelled over the ground in gliding is smaller whether the machine is brought down faster and steeper or slower and flatter than the most efficient gliding speed and angle.

he decreased his speed by raising the nose and gliding below his most efficient speed. Sometimes, when in doubt, it may be wiser and safer to come down on the fast side and then hold the machine off the ground as long as possible, rather than to come down on the slow side with the possibility of stalling on the descent, especially when turning. This applies particularly to getting off followed by an engine failure near the ground. To turn back and land in the aerodrome necessitates magnificent judgment and skill, as the pilot will otherwise certainly stall, sideslip and crash. Therefore, the rule is that, in 99 cases out of 100, it is better to go on and land in a straight glide even on bad ground. Hundreds of pilots, even of great skill and experience, have crashed through failing to observe this golden rule. In fact, the more experienced a pilot is the more is he likely to think that he is able to perform the extremely difficult manœuvre of making an about-turn down wind from a low height. Sometimes it is a wise plan, when an engine gives out over bad country, to turn down wind for the first few thousand feet of the gliding descent with a view to reaching better country, the length of glide down wind being, of course, very much farther than in any other direction. The pupil should then turn into the wind at a height of about 1000 ft. or 1500 ft., in order to make certain of his field. However, he must not rely too completely on his altimetre, because the ground under him may be several hundred feet higher or lower than the spot where he last set his altimetre to zero. If he is very high up he can judge the direction of the wind by making a spiral and watching the way it blows him. By studying all these signs continually, he has no hesitation in which direction he ought to land.

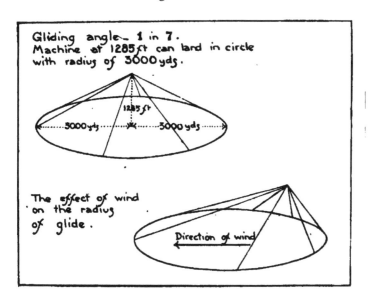

Top: Illustrating the radius of glide of a machine in still air. Bottom: Showing how wind affects the radius of glide, which is greater down wind, but less up wind. In each case the machine is supposed to be at the apex of the cone of glide when the engine is throttled down.

Selecting Suitable Landing Ground

The moment that his engine stops he must put the nose of his machine down and begin to look for a suitable field, or, better still, a group of fields where he may land. A group of fields is chosen in preference to a single field because he may easily overshoot or undershoot the mark, and it is therefore just as well to have alternative landing places. For preference, a grass field should be chosen possessing length in the direction of the wind, and without such obstacles as trees or houses surrounding it. A grass or stubble field, or a sandy beach makes the best landing grounds. Cultivated areas are generally comparatively level. Grass fields may have sheep or cattle on them and, naturally, look green from the air. Stubble is dark brown,

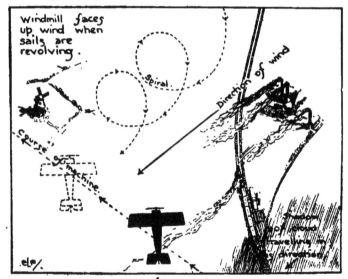

Methods of judging the direction of the wind from the air. (1) By noting windmills. (2) By watching the drift of the machine in a spiral. If there is no wind, the machine would spiral round a fixed point on the ground. (3) Smoke from chimneys, trains and rubbish heaps. ((4) The drift of the machine near the ground. (5) The shadows of clouds.

a ploughed field looks black, and roots dark green. Golf courses can be distinguished by the yellow or brown patches formed by the bunkers. Plough, crops or roots are all bad, and only a pancake landing will prevent the average pilot from turning over. The golf course may be chosen if nothing better is at hand, and the same remark applies to parks and public recreation grounds. As roads and railways are generally bordered by telegraph wires, it is a bad plan to land near them, as the wires cannot be seen until the machine is upon them.

Effect of Snow

If the pilot is flying when there is snow on the ground it is practically impossible to pick out the kind of surface field desired. He can console himself with the fact that all the ground will probably be frozen hard, and the nature of the terrain is not of such importance as it would be under more ordinary conditions. He will find, however, that he will have to be more careful than usual in landing, as snow is very deceptive owing to the different lighting of the ground that it causes. Incidentally, he will find it much more difficult to recognise landmarks after a heavy fall of snow, as roads, railways, fields and villages are all included in one great camouflage. The same remark applies to extensive floods, which blot out distinctive bends in rivers that could be used to give the pilot his bearings.

The lighting of the ground varies enormously if the pupil be flying late in the evening while the sun is setting; he will fly in its rays, according to its height, after it has disappeared from the view of those on the ground. When he. comes to earth he will suddenly note the contrast in the lighting near the ground, where it is so much darker than it was when seen from high up, and for this reason he would do well to circle once around the aerodrome to accustom himself to the comparative dullness.

Judging Distance

Rules for judging distance are as follows:—A pilot is inclined to over-estimate distance in bad light, in mist, and when the field and background are of the same dull colour. He will under-estimate distance when the light is very good, or when he is landing with the sun behind him.

Another difficulty when landing is to tell from above whether the ground is level or not. This can only be seen when within 200 ft. or 300 ft. of it. Landing uphill means that the machine must be flattened out to the stalling point with its tail well down. A downhill landing, which should be avoided if possible, means a very much longer travel before the machine comes to rest. It is a wise plan always to land the long way of a field, or, in some cases, to make the diagonal from corner to corner, as the extra bit of ground gained may make all the difference between a crash and a safe landing.

In the event of there being no field of ideal dimensions at hand, it may be necessary for a pilot to make a landing across wind. This is quite easily done, although it is not often practised in school flying. The method of doing this is described on page 159.

Making an "S" Turn

In any case, the procedure for hitting off the field, whether it lie into the wind or across the wind, is the same. The aim of the pilot should be just to glide over the hedge or ditch on the near side of the field. To do this he will find that a process called "S" turning is most useful. This consists of making a series of "Figure 8" or "S" turns as he descends. By this method the pilot can always keep his eye on the objective, if necessary making several "S" turns over the same ground. In making these turns, he must always turn towards the landing ground, and during the descent he should study the field most carefully, noting the position of trees and watching for any obstacle on the ground, such as agricultural implements and heaps of rubbish. If a spiral were attempted instead it would be quite possible for the pilot to find himself half a spiral turn out, that is, facing away from the field, at the moment that he wanted to make his final glide into it from the last 200 ft. or 300 ft. "S" turns can be practised on any aerodrome, and they will be found very useful for future emergencies. As the ground is approached they can be made shorter and shorter until there is no doubt that the field will be made safely.

Another good method of bringing off a forced landing is to make a wide sweep into the field before turning into it. If the machine were falling short owing to the distance having been misjudged, due to the strength of the wind, the pilot can turn in quickly and much earlier than he would if he wanted to use up space. The latter would be the case if he were overshooting the mark.

Another method used at times is to select a group of fields as before and to allow the machine to glide naturally towards it. When nearing the ground the pilot will have an excellent idea as to which field the machine is likely to hit, and, with a little control, he can help it to attain this end. The illustrations depict the various methods described.

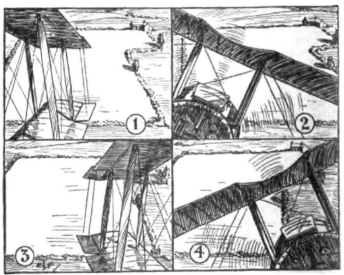

Showing how a pilot, by making "S" turns, can always keep his eye on his objective, i. e., the field he intends to land in. (1) Right bank turning towards the field. (2) Nearly round. (3) Left bank turning towards field. (4) The left-about turn nearly completed.

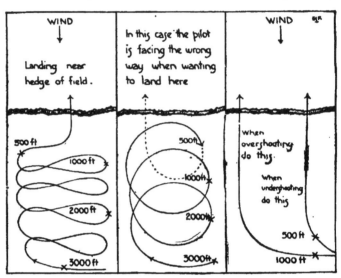

WIND

Landing near hedge of field.

In this case the pilot is facing the wrong way when wanting to land here

WIND

500 ft
1000 ft
2000 ft
3000 ft

500 ft
1000 ft
2000 ft
3000 ft

When overshooting do this.

When undershooting do this

500 ft
1000 ft

Points to bear in mind when landing in fields.

Turning Near the Ground

In turning near the ground—an unwise procedure at any time—great care must be taken to maintain the speed of the machine by keeping the nose down. Hundreds of accidents have occurred through pilots stalling their machines by turning too flat when trying to make a forced landing and endeavouring to lengthen out their glide, or to avoid some hitherto unseen obstacle. At the same time, it is a good plan to land as slowly as possible in order to shorten the carry of the machine. A fast landing means a long run and the possibility of the machine hitting the hedge at the other end of the field. In order to avoid such an accident, the pilot should turn the machine away to one side while it is running along the ground after landing, but this should not be attempted when the machine has a good deal of way on it. A pancake landing can sometimes be made intentionally on bad ground. To stop a machine on the ground the pilot must get the control

lever back to his chest, which brings the tail down and
the tail skids into contact with the ground, where their
added weight and friction soon slows up the machine. If
this is done too soon, however, the machine may display
a tendency to jump into the air again, and if it does this
it may even pancake heavily. Pupils should remember
to land slowly at all times, and especially is this necessary in hay, corn, plough, snow and marsh. Here, again,
a pancake landing is the safest method. Frozen ground
is generally safe, but it must be remembered that a machine runs a very long way on it, and due allowance
must be made for this. Incidentally, frozen ground is very
much harder on a machine than thawed earth. The difference can actually be felt after a frost, especially when
giving landing practice.

Steps to Take After a Forced Landing

Assuming that the pilot has come down successfully—a
creditable piece of flying, for even the best pilot may
crash on a forced landing, because, even from a height
of 1000 ft. or so, one cannot make absolutely certain that
the field chosen is suitable, and he cannot see wire fencing,
small dips or hillocks, ditches overgrown with grass, ridges
and ruts, heaps of stones, mole hills, and even such things
as small agricultural implements, any of which may be
enough to overturn the machine—after landing, he should
ascertain first of all what is the matter with his engine,
and then telephone or wire to the nearest aerodrome or
headquarters stating his whereabouts and describing the
trouble. In stating where he is he should be able to say
if there is a road near by suitable for the tender to use if
one is being sent to salve the machine. When a pupil
wishes to describe his exact position on a map, he can
either give his bearings and distance from the nearest town,
or from some obvious landmark, or he can report that he
has landed on some letter in some place printed on the
map, giving at the same time the map number and its name.
Thus, if he said that he was down at the third "s" in "Sus-
sex," or the "y" in "Canterbury," on Sheet 12 of Bacon's
2-ins.-to-the-mile road map, his commanding officer, having
a similar map in front of him, would know his whereabouts

exactly, and would instruct the breakdown gang accordingly. If he has damaged his machine in landing, he must
inform his headquarters of all the damage. If he mini-
mises this, it is possible that the tender and the mechanics
who may be sent out to his rescue with spare parts, owing
to the insufficient information supplied, may not have
brought all the necessary spares, and a second journey may
have to be made; hence the need for giving the very fullest
details of the damaged parts and spares required. To do
this, a pupil should know the correct technical terms for
the various parts of an aeroplane and its engine. If the
engine has failed it is always much more creditable if
the pupil can diagnose the trouble and repair it himself.
A broken petrol pipe, faulty switch or contact breaker, a
sooted sparking plug, or even a broken valve can be replaced with the aid of a few simple tools, and such operations are worth attempting by a pupil possessing a certain
amount of mechanical knowledge. If the machine is
damaged, the pilot must take the necessary steps to secure
it for the night. He must obtain assistance and move the
machine, if this is possible, under the lea of a hedge, trees,
haystack or house. The machine should be pegged down;
that is, pegs are driven into the ground at each side of the
tail and a little in front of each wing tip, taking care
to insulate the rope from the wings and tail with sacking,
straw or other material at hand.

Pegging Down a Machine

The machine is tied up to the pegs, which must be driven
deeply into the ground in case a strong wind should spring
up in the night and blow the machine away. Some slack
must be left in the rope or there will not be sufficient play
to allow the machine to rock slightly in the wind. Some
machines are provided with rings under the outer struts
for the purpose of being pegged down. It is then wise
to cover the propeller, engine, pilot's seat and instruments
with tarpaulins or sacking in order to keep out the damp.
Sometimes it is as well to remove such instruments as the
watch and also the tools from the machine. If the machine is to be left out in a very strong wind, ruts may be

How to peg down a machine for the night. Pegs are driven into the ground at the wing tips and on each side of the tail. Wings and tail are secured by rope, but not stretched too tightly to the pegs. The controls should be lashed central, and if there is a wind the machine is left in the lea of a hedge or house. The propeller and pilot's seat can be covered with sacking or oilskin.

dug to accommodate the wheels, the tail can be jacked up, and the elevator control lashed back so as to present as little incidence to the wind as possible and thus to prevent the wind overturning the machine. Additional ropes can be slung round the propeller.

Having secured the machine and made arrangements for a guard—in England local or military authorities or special constables will undertake the work, and in France the civil or military authorities, such as the mayor of the village or the commandant—the pilot can depart to his temporary headquarters to await orders or the breakdown gang. He should inform the guard and also his C.O. as to where he can be found, or at what time he will be at such-and-such a rendezvous. If possible, he should give the number of the telephone or the name of the local post office or hotel.

Special precautions necessary to peg down a machine in the open in a strong wind. Note trenches dug for wheels and machine trestled up in the flying position, so as to reduce the lift on the wings.

Starting Up Without Aid

If he has been able to repair the damage himself, he will be faced with the problem of restarting his engine. He must obtain chocks—bricks, fence poles, or blocks of wood being used for this purpose. If he cannot obtain such things, he can get the observers—and there are certain to be plenty of these about—to hold back the machine while he tests the engine. He should explain to them where they can lay hold of the machine, i. e., by means of the struts and as near as possible to their lower sockets.

Engine Starting Hints

Some difficulty may be experienced in starting the engine. On a machine with a stationary engine which will throttle down well, the pilot can start it himself. The explosive mixture is sucked in with the switch off, and possibly a rag over the air intakes to the carburetter. He then walks round to the switch, putting this to contact, closing the throttle so as to allow the engine to fire at a very low number of revolutions; he can then start the engine in the ordinary way by swinging the propeller. This operation presents no difficulties whatever, for the pilot should have practised propeller swinging during his early flying experience.

Swinging a Propeller

Perhaps a word on swinging propellers may be of value at this point. Practise first with the engine off. The hands should be laid on the propeller about 4 ins. apart and about one-third of the distance from the tip of the propeller. The fingers should not hook over the trailing edge ; a ring is liable to catch on this if the engine backfires, and should therefore be removed. It is not advisable to stand on muddy ground, as a slip might cause an accident. When the switch is put on contact the propeller should be swung fearlessly, making sure that the engine is taken over compression. If this is done there is very little chance of the engine backfiring, which it is far more likely to do if the propeller is swung in a timid and half-hearted manner. Generally, the throttle should be partially open : the more the throttle is open the greater the compression of the engine to be overcome. It is a bad plan to stand at any time in line with the propeller, as, for instance, when holding chocks. Some pupils have difficulty in finding out which way the propeller revolves, but if they think out how it is designed with a view to cutting its way through the air, they should be able to distinguish the entering edge of the blade from the trailing.

Starting a Rotary Engine

On rotary-engined machines the method of starting up is not so simple. With local help, one of the onlookers may be taught to swing the propeller, or, alternatively, to switch the engine on and off as soon as it fires. There is, however, the chance that the inexperienced and probably highly-nervous farm hand, when in the pilot's seat, may lose his head, and, if there are no chocks, may run down the pilot, who is in the front of the engine after swinging the propeller. Therefore, a tip is given to provide against this mishap. Tie the switch to a peg in the ground behind the machine, so that if the machine begins to move the switch will be pulled off by the peg. It may be mentioned here that the system on which a magneto switch works is to short-circuit the current, i.e., to make or complete the circuit. This is quite the opposite to a switch working in conjunction with accumulator ignition, where a switch is on when a circuit is broken and off when a circuit is made.

Importance of Clean Wheels

If the pilot finds that he has landed in a very heavy stubble field, or in thick mud, he should see that his wheels are clean before he restarts, as otherwise the mud

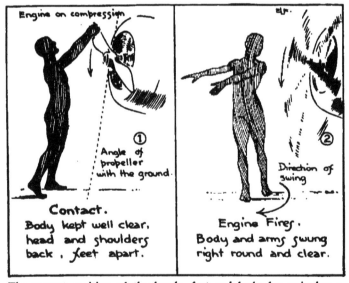

Engine on compression

① Angle of propeller with the ground.

Contact.
Body kept well clear, head and shoulders back, feet apart.

air.

② Direction of swing

Engine Fires.
Body and arms swung right round and clear.

The correct position of the hands, feet and body for swinging a propeller.

and dirt will collect on them and will then be flung up on to the propeller by centrifugal force, which may possibly break or chip it. This is indicated by the whistling noise which a chipped propeller makes in the air. If a machine becomes bogged, assistance may be obtained and men sta-

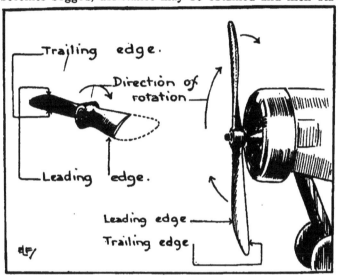

Trailing edge.

Direction of rotation

Leading edge.

Leading edge

Trailing edge

Showing which way the propeller revolves, a point of difficulty with some pupils.

tioned under the wings to lift the weight off the wheels whilst the pilot races the engine to move the machine forward. The pilot must be careful not to overrun the men supporting the wings.

Causes of Forced Landings

The common cause for a forced landing is running out of petrol, possibly due to starting with the tanks nearly empty, to a leak, or to a pilot forgetting to keep up the pressure in the tank on a machine with pressure-feed supply. Some engines require doping or priming with petrol in the cylinders before starting. A dope-can, which is a small petrol squirt, should be carried on such machines when proceeding on cross-country flights; but if this is not at hand the engine can be primed by means of a rubber tube or a bicycle pump, or, in extremis, by soaking a handkerchief in petrol flooded from the carburetter and

How to taxy a machine out of heavy ground. The pilot races the engine while the men lift under the main spars of the wings. Planks can be placed under the wheels.

squeezing out the contents through the open exhaust valve or sparking plug orifice. A petrol funnel can even be made out of a piece of paper, and petrol poured through it into the combustion chamber.

Warming the Carburetter

In cold weather, or when a machine has been out all night in the open, difficulty may be experienced in restarting the engine. It should be noted that water-cooled engines should have the water drained off from all points before being left out on a cold night, or the water may freeze, unless a non-freezing solution of glycerine is used. In restarting, hot water should be put into the radiator, care

being taken to see that all taps are turned off first.
hot rag placed round the carburetter and induction p
will help to vaporise the fuel. For easy starting t
engine may be primed and the sparking plug points plac
close together, but not touching, although sooting up
more likely to occur under these conditions. In some e
gines one particular plug may be liable to soot up. In th
case it may be given an extra wide spark gap, the id
being that the engine can be started up on any of the oth
plugs, the cylinder with the big spark gap plug chippi
in when the revolutions of the engine increase and th
strength of the electric discharge at the plug jumps th
extra wide gap.

Causes of Engine Failure

Another frequent cause of trouble in starting is leak
induction pipes. A useful method of preventing leaks a
such points is to wrap up the induction piping, especially
the joints and unions, with insulating rubber or medical
tape, which will prevent the ingress of air and hence
keep the mixture correct.

A broken petrol pipe can be repaired with rubber tubing, a piece of which, the same size as the petrol supply
pipe, should always be carried. Insulating rubber tape
may be bound over the rubber joint and copper wire over
the top, so as to make all secure.

It has already been said that the engine may show signs
of giving up long before it fails altogether. This may be
due to faulty lubrication, or, on a Gnome, to a faulty or
worn obturator ring, which can easily be discovered by the
blueing-up of the top steel radiating fins of the cylinders.
A faulty petrol feed will cause a drop in power, probably
accompanied by spluttering noises. Generally this will be
due to the perishing of the india-rubber or composition used
in the petrol pipe, which, accordingly, should be inspected
from time to time, and renewed if there are any signs
of the composition perishing or crinkling up on the inside.
A choked jet is a more rare occurrence, but it is wise to
see that the filter and the gauze below the carburetter
are cleaned occasionally. The same remark applies to the
oil sump and strainer found on many engines. These

first should be cleaned with paraffin occasionally. They must be taken apart and replaced with fresh gauzes if the old ones are cracked or broken.

A point worth noting when flying dual-control machines solo is to see that there is nothing in the other seat to jam the control levers or rudder. Cushions should be strapped

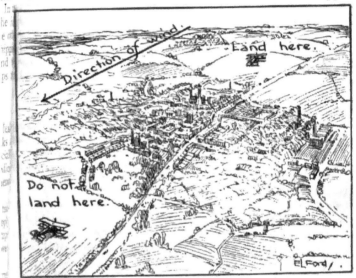

An example of good and bad airmanship. The pilot should land to the windward side of the town, so that in getting off again he has clear country in front of him.

down. The belt should be fastened across the seat, and a suit case, if carried, should be roped up in such a manner that it cannot foul the controls.

Good and Bad Airmanship

There are many other tips which the pupil-pilot will only learn by experience, but the probability is that, in the course of events, he will complete his journey successfully. An example of the kind of experience referred to may be given. The pilot has to land close to one or other side of a large town, the wind blowing from the east across the town. If he landed on the east side of the town, where the landing ground looked good, he would have to get off over the town, with a consequent chance of making a forced landing in the town itself if his engine failed; so he lands to the west of the town instead, and thus has a clear line of country in front of him when he starts again.

A Landing and a Lesson ·

Another case may be quoted where an experienced pilot would have saved his machine, whereas an inexperienced pilot wrecked his. In the course of a flight in the neighbourhood of an aerodrome, the switch on a Gnome engine machine failed, and the engine could not be switched off when the pilot desired to land. The pilot should have made certain of getting within gliding distance of the aerodrome and then turned off the petrol and glided home. Instead, he attempted to govern his engine on the petrol supply, with the result that he stopped it. He then saw that he would not reach the aerodrome, as the length of the glide of the machine, plus the amount that it would have been assisted by the engine being switched on and off in descending, was obviously shorter than the glide of the machine without the engine, so the pupil had to make a forced landing in some rough shingle, which he brought off successfully by pancaking. There was a strong wind blowing at the time, and in turning round, tail to wind, in order to get a long run to take off, he forgot the softness of the shingle and used too much engine, with the result that he attained too much speed, and, with his control lever too far forward, allowed the wheels to sink into the shingle, whereupon the wind got under his tail and slowly turned the machine over on its nose. It had to be dismantled and taken back to the aerodrome in a tender.

A point to bear in mind when taxying down wind is that it is well to keep the elevator slightly down so that the wind keeps the tail low.

Another instance was that of the pilot whose throttle control lever broke. As is the case generally, the carburetter control was set so that, in the event of the control breaking, a spring would hold it wide open; thus he still had his engine, although it was running all the time at maximum revolutions. Now the pilot knew that if he ran his engine for long at this speed it would almost certainly break down, so he adopted the ingenious plan of climbing the machine at maximum power and steepness to a great height, and then switching off the engine altogether and making a long, slow glide down again. He repeated this process again and again, and eventually reached his destination safely. A less-experienced pilot would either have run his engine all out all the time, or attempted to govern it on the switch, both of which methods might have ended disastrously. In the case quoted, the engine, when climbing on full load, naturally slowed down considerably, and so avoided the critical number of revolutions.

To restart the engine, should it stop when switched off, and there not being sufficient gliding speed to keep it going, it is only necessary to dive the machine steeply,

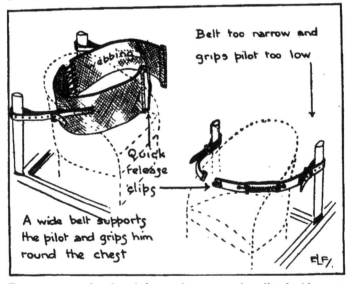

Two patterns of safety belts used to strap the pilot in his seat. The wide belt is far the better of the two.

when the force of air passing the propeller will generally restart the engine.

These examples might be multiplied, but they serve to show the kind of thing that makes the difference between good and bad airmanship.

The Homeward Journey

In landing at "A" aerodrome, a pupil should remember to face the wind and land towards the head of the arrow if one is used. Sometimes a smoke flare is used to indicate the direction of the wind, in which case the pilot must land parallel to the smoke and towards the bucket from which the smoke is coming. Sometimes a kind of sausage balloon is fitted on the top of a pole for the same purpose. He should then taxy the machine up to the hangars and report himself to the duty officer and to the commanding officer, if necessary. Before returning, he should make certain that his petrol and oil tanks are sufficiently full, and take all the necessary precautions and preparations for the return journey that he did for the outward.

If the pupil is returning late and is in danger of being overtaken by darkness, or if he loses himself late in the evening, he must remember to begin looking out for a suitable landing ground before it is too late. Darkness may come upon him, if he is flying high, quite suddenly, so that he must be on his guard against this and come down and land when the conditions of light are still favourable.

It is a bad thing to attempt a long cross-country flight on an empty stomach, and this is especially the case in early-morning flying, or when the air is very cold. A good meal or a hot drink should be taken even before a short ante-breakfast flight.

Chocolate, tablet food or biscuits are good for taking during a flight, and will keep a pilot going for a long time; but it must be remembered that it is none too easy to dissect tablets from other articles and to withdraw them from the pocket. The best method would be to arrange them in a rack or ledged shelf in front of the pilot on the instrument board.

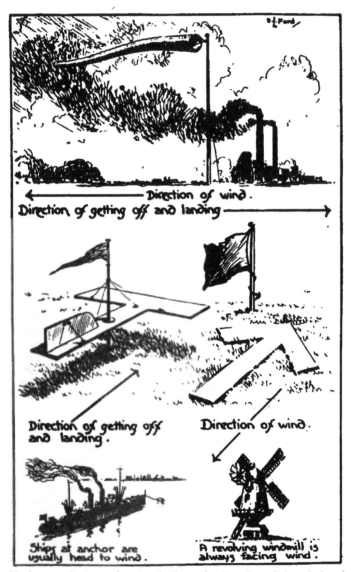

Various methods employed to indicate the direction of the wind and other signs which can be observed by the pilot, who, during all his cross-country flights, should pay particular attention and notice the direction of the wind.

185

CHAPTER X

More Advanced Flying

WHEN a pilot has reached a certain degree of proficiency, in all probability he will wish to practise fancy flying, such as, on occasion, he has seen more expert aviators perform. Although this fancy flying, or stunt flying as it is called, is excellent training for aviators who are intended for scout or war flying, it should be undertaken with caution and at a fair height. "Stunting" near the ground is at all times dangerous, and many good pilots have crashed through overlooking this warning.

An Ascent of 10,000 Feet

Any pilot who is to attain fair proficiency should be able to make an ascent of 10,000 ft. while still in the pupil stage. There is nothing difficult or dangerous in such a climb. It merely requires patience and a certain amount of nerve to carry it through for the first time. Pilots who are unaccustomed to long climbs are apt to climb their machines less and less steeply as they get higher, under the mistaken impression that they are stalling, or inclined to stall, their machines. It is true that, on the other hand, they may be able to get better results from their machines by climbing them slightly in excess of the lowest flying speed. This can only be tested by the watch.

It may be mentioned here, perhaps, before going on to describe some of the fancy tricks that may be attempted at the higher altitudes, that when a pupil first goes up really high, i.e., in the region of 10,000 ft. or more, generally before landing again he should fly once round the aerodrome at 1000 ft., so as to accustom himself again to the appearance of the ground at this height. Owing to the difference in air pressure at various heights, the pilot will do well to descend slowly, and if he notices any strange feeling in his head or in breathing, to swallow now and then so as to equalise the pressure of the air inside his head with that outside.

Less Petrol for High Flying

As the atmosphere is rarer the higher he flies, he may find that he has to fly at a faster speed in order to get the best results.   This is due to the pressure of air on the air speed indicator being less at heights for the same speed than lower down and nearer the ground.   For the same reason he may notice that the carburation of the engine varies considerably, and he may find, too, that he will have to cut his petrol down in order to counteract the effect of the smaller percentage of oxygen in the air at considerable heights.   This can be more easily done on a machine with a Gnome type engine, in which the petrol is controlled by a needle valve.   If a hand control extra air valve is fitted, this can be opened to obtain a correct mixture.   When descending from a height, the engine can be cut off altogether, to be started again at 1500 ft. or 1000 ft. by simply diving the machine, which allows the pressure of air again to revolve the propeller.   As pointed out previously, when descending with a Gnome engine machine, it is wise to keep the switch on and turn off the petrol to prevent the sparking plugs becoming oiled up by superfluous oil flowing about the combustion chamber instead of being used up as it would be when running in the ordinary manner.   On tractor machines it is quite difficult to stop the propeller in the air if any speed is attained; but in pushers the reverse is the case, and it is next to impossible to restart the propeller once it has ceased to revolve.

A Vertical Bank

The high climb to 10,000 ft. which has just been described will be the first advanced test that the pupil will attempt.   He can next try a vertical bank—a term applied loosely to all banks of 45 degrees or over.   There is no

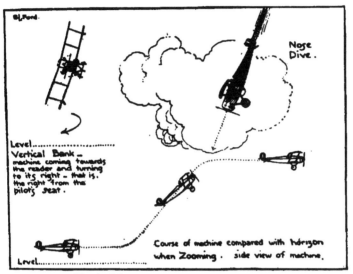

Vertical bank, a nose dive and zooming.

particular difficulty in this manœuvre: it is simply an exaggerated bank, and is performed by putting the control lever and rudder farther over than in the case of an ordinary turn. As the aeroplane is now on its side, the nose of the machine must be trimmed to the horizon by the use of the rudder. Top rudder will make the nose rise in relation to the horizon, and bottom rudder will make it descend. To make the machine turn, the control lever is pulled back and slightly over in the opposite direction to ease off the bank, because, otherwise, the bank would keep on increasing owing to the greater effect being exerted by the aileron on the outside wing which is moving so much faster than the inner wing. Thus the correct procedure, in order to make a vertical bank, is to put the rudder and control lever over in the desired direction, then to trim the nose of the machine to the horizon, in the case of a level turn, with the rudder over (or below the horizon if it is a gliding turn), and, as this is done, to ease the control lever back to make the machine turn,

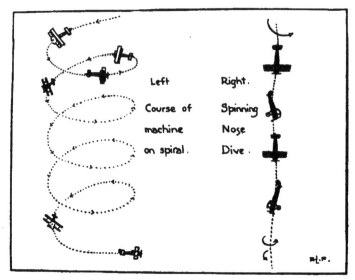

The course followed by the machine in making a spiral descent,
and what happens in a spinning nose dive.

and slightly in the opposite direction to prevent the bank
increasing. Now, to come out of the turn, it is necessary first to give the full opposite bank, and then, when
the machine is nearly horizontal again, to give any opposite rudder that is necessary and to put the control lever
central again and slightly forward. If the opposite rudder were given before the machine were nearly horizontal,
the effect would be to make the nose rise and a stall might
result. If he continues his vertical bank round and com-
pletes the circle, he is in a fair way to attempt a spiral,
for a spiral is a circular vertical bank made with the engine cut off.

Spiral Descents

In making a spiral the pilot should always look downwards and inwards towards the centre of the circle that he
is making. If he finds his angle becoming too steep, he
merely takes off a little bank, or pulls the control lever

back slightly if his descent is too rapid. He can judge of the accuracy of his spiral by watching his string or tape, the operation of which was described in the chapter on the instruments. If the spiral is correct, the tape should trail back in a perfect line with the fuselage of the machine. When a bank or spiral becomes extremely steep, a curious phenomenon occurs: the rudder acts as an elevator and the elevator as the rudder. This is a somewhat loose way of expressing what happens, for if one regards the rudder as a controlling surface operating the machine in a certain plane, it will always control the machine in this plane, regardless of its position relative to the earth. The elevator may be regarded in the same light and controls the machine in the same way.

Inversion of Controls

This inversion of controls is really due to the different positions assumed by the nose of the machine relative to the horizon. When one is flying level, a backward movement of the control lever makes the nose of the machine rise above the horizon; but if the machine is on its side, as it would be in a vertical turn, the same movement of the control lever makes the machine turn more quickly. In each case the nose of the machine is trying to reach the tail—an extreme instance of this being the loop. In the same way, left rudder will always make the nose of the machine try and meet the left wing tip, but if the machine is on its side, say, in a left-hand turn, left rudder will make the nose of the machine drop below the horizon, although it is still trying to reach the left wing tip, which is, of course, underneath.

"Zooming"

The next trick he can practise is the "zoom," or the sudden jump of the machine several hundred feet into the air after flying near the ground. First of all he must remember that he cannot effect a zoom until he has got up full speed, for it is only the surplus speed that allows the machine to climb so steeply and suddenly. The manœuvre is performed by pulling the control lever back suddenly, which causes the machine to climb very quickly and steeply, and then putting it forward again when the machine has practically reached the stalling speed, as indicated either by the air speed indicator (which is a bad sign owing to its amount of lag) or the general "sloppy" feeling of the controls. When a pilot notices this peculiarity at any time in the air he must beware, because it means that the machine has lost its flying speed and will stall the next moment unless he pushes the control lever forward and allows the machine to regain its velocity.

Nose Diving

The nose dive is the opposite to the zoom, and is performed by putting the control lever forward. It is wise to cut off the engine before making a nose dive, because this minimizes the strain on the whole machine. Nose dives should not be attempted too near the ground, and a pilot should not attempt too steep a dive to start with. He should see, too, that his belt is tight, as otherwise he may notice a tendency to slip through it on to his controls and instruments. He should also make certain that his goggles fit well, as if there are any air leaks in them they will cause his eyes to water and consequently blur his vision. He will do well to bring the machine out of the dive at over 1000 ft., in order to allow plenty of room for eventualities. To do this, he should pull the control lever back firmly but not too rapidly. He will feel the machine levelling up, and as the air speed indicator does not register as quickly as the machine changes position, he must not centre the stick until the instrument shows him that he is somewhere near his lowest flying speed. A nose dive can easily be followed by a zoom if the pilot pulls the control lever back rapidly.

Looping

Should he do this, and should he, in addition, put his engine on full at the same time as he pulls back his stick to the limit, he will, in all probability, loop, although

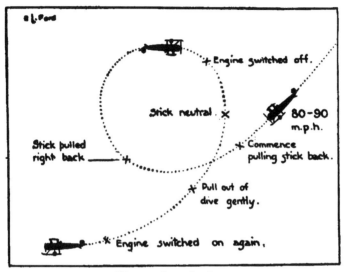

Stick neutral . ✗

Engine switched off.

80-90 m.p.h.

Stick pulled right back ___

Commence pulling stick back.

Pull out of dive gently.

Engine switched on again.

How to loop the loop.

this is not by any means the safest or best method of
performing this very simple but, at the same time, very
effective manœuvre. A pupil who wants to loop should
select a machine that is known to loop easily, such as an
Avro or B.E.2c. He then ascends to a height of 3000
ft. or 4000 ft., gradually puts the nose of the machine down
to a speed of 80 m.p.h. or 85 m.p.h. (75 knots or 80
knots), attaining this maximum by a more gentle descent
than would be possible if he attempted to nose dive to this
speed. He then pulls the control lever as far back as it
will go, in one firm, strong pull, the effect of which is to
cause the machine to rear vertically upwards and over.
When he is upside down, he will see the ground below him
and must then cut off his engine and a few moments later
ease the stick, gradually centring it. The engine can be
switched on again when the steepness of the nose dive has
been materially decreased. The first part of the pull-back
should be slower than the latter, on account of the greater
speed of the machine in the early stages of the loop. The

control must be held back until the machine has completed
the loop.

Before looping, the pilot should see that the machine
is perfectly trued up and that all the wires are correctly
adjusted. He should also make sure that his belt is
strong enough to hold him, in case he should make a
bad loop or stall in a more or less upside-down position.
If he did a correct loop there would be no need for him
to wear a belt, as the centrifugal force of the manœuvre
would pin him to his seat. It is dangerous to have anything loose in the front seat of the machine, and such
things as cushions should be removed before attempting
any of these evolutions. In looping a machine like the
B.E.2c. the pilot must brace himself to keep the rudder bar
level, and also to pull the control lever back straight and
not to one side, or else he may find that he will not come
out of the loop square. In looping some machines in

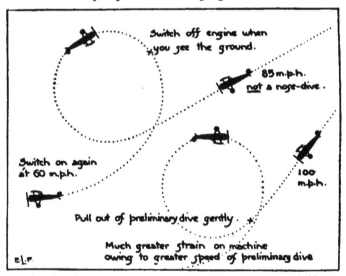

Showing the right and wrong way of looping. It is a much bigger
strain on the machine to pull it up for the loop after a steep dive
than after a gentle one, where the necessary speed is attained
gradually.

which a rotary engine is fitted it is necessary to use rudder to counteract the effect of engine torque when the engine is cut out, consequently it is advisable to practise on a simple type of machine first. Looping is generally done into the wind.

Tail Sliding and Spinning

Another favourite trick of experienced aviators is to stall, or tail slide, the machine at a fair height and allow it to flutter down some distance. This is done usually by cutting out the engine and pulling the control lever

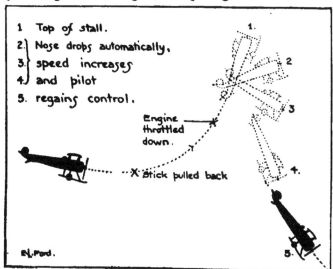

1 Top of stall.
2. Nose drops automatically,
3. speed increases
4. and pilot
5. regains control.

Engine throttled down.

X Stick pulled back

R.Ford.

The effect of stalling the machine, a perfectly safe and easy manœuvre given the necessary height.

back. The machine then loses its flying speed and falls over on one wing or slides back tail first. If the nose is put down or the engine restarted, the machine will gather speed again and the pilot will regain control. If the engine is cut off and the pilot holds the control lever back all the time, he will come down in a series of stalls and their ensuing dives.

A tail spin, sometimes performed unintentionally by inexperienced pilots, is another stunt practised by skilled aviators. Sometimes it is achieved in conjunction with a nose dive, in which case the evolution goes by the name of the spinning nose dive or corkscrew spiral. What happens in this case is that the pilot stalls his machine, pulls the control lever towards him and fully back. He does this

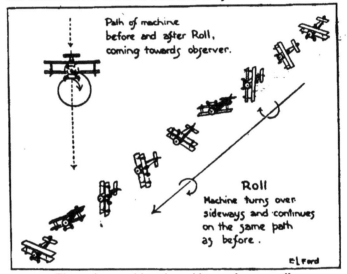

Path of machine before and after Roll, coming towards observer.

Roll
Machine turns over sideways and continues on the same path as before.

El Ford

The various positions in a sideways loop or roll.

with the engine off, and then rudders hard in the direction in which he desires to spin. He can get out of the spin, if he has sufficient height, by placing the controls—rudder and stick—central, whereupon the machine will take on a nose dive, when the engine can be restarted and the flight continued in the ordinary manner. Pupils who find themselves in a spin unintentionally must remember this. It is generally found that a pupil will get into a spinning nose dive through making a faulty spiral on certain types of machine on which the area of the stabilising tail fin is too small.

In stunting rotary-engined machines the gyroscopic effect

of the engine, which causes the nose to go up if the machine is turned in the direction of engine rotation, and vice versa, will be noticed.

The roll, which consists of making the machine loop sideways and continue in the same direction as it was travelling before the manœuvre, is done with the engine on or off. The pilot puts on speed and then pulls the control lever back as in a spin, and kicks on full rudder in the same direction in which he wishes to roll. To sideslip—which means that the machine descends sideways much more rapidly than it goes forward—full bank is given in the direction in which it is desired to slip whilst rudder is held off.

Rolling and Staggering

An evolution that is very easy to perform consists of making the machine take on a rolling and pitching move-

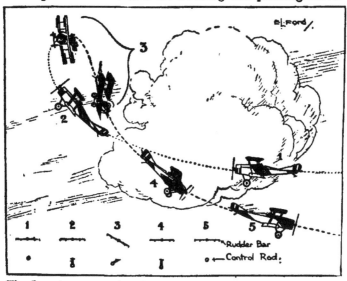

The Immelman turn, in which the machine rears up, turns sideways over the vertical, and comes out facing in the opposite direction. Below are shown the movements of the control lever and rudder bar required in making this manœuvre.

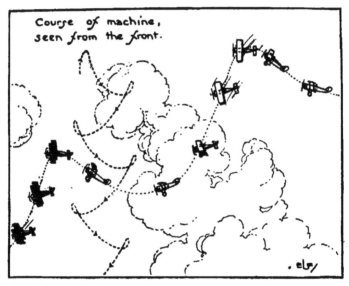

Courfe of machine,
seen from the front.

The falling leaf descent and engine cut-out, seen from the front
and sideways.

ment by working the control lever round and round in a
circle and ruddering accordingly, i.e., giving a left rudder
when the control lever comes round to the left, and right
rudder when it comes round to the right. The effect on
the machine is to make it stagger and see-saw like a drunk-
en man.

"Cartwheel," "Immelman Turn," "Boot-lacing" and "Falling-leaf" Stunts

Another variety of stunt is termed the cartwheel, which
can be mistaken for a loop, in certain circumstances, by
spectators on the ground. It is performed by getting up a
little speed, by putting the control lever forward and then
pulling it back, as in a zoom. When the machine is almost
standing on its tail, but before it has lost flying speed and
controllability, apply rudder and bank in the same direc-

tion. The machine will answer to the controls, cartwheel in the air, and come out facing in the opposite direction. A slight modification of this manœuvre results in the famous Immelman turn. The engine can be cut out when the machine turns about, and will allow it to dive, but if this stick is held fully back the machine will come out of the dive quite easily. This manœuvre can be done with the engine off, the necessary momentum for the ascent and cartwheel turn being supplied by diving. When the machine is pulled over, first on one side and then on the other, in the course of a long descent, the manœuvre is sometimes called "Boot-lacing." Stalling turns are very similar to these manœuvres, except that all way is lost at the top of the pull-back, and the machine then drops its nose suddenly and falls over on one wing tip, and so comes under control again. The "Falling-leaf" descent is a modification of this manœuvre.

On some machines fitted with large rudders it is possible to descend with the control lever right back. A very slow stalling descent, practically vertical and flat, results, although speed must be increased to normal in order to land. It is not easy to keep the machine from falling over on to one wing or the other during this manœuvre, but it can be done by working the rudder more roughly than ordinarily, so as to hold the wings up.

A Flat Turn

A very quick method of turning is attained by switching off the engine momentarily and kicking on full rudder, and then centring it. The machine will drop most of its flying speed and turn round through about 90 degrees. Indeed, it would continue round and get into a spin if the control were not centred again. In a spin on some machines, the lever must be pushed forward beyond the centre position. Upside-down spinning has also been performed, in which case the control is reversed, as it would be in upside-down flying, which has been done on certain types of Service machines.

In all these evolutions it is a common occurrence for a pilot to get into the "wash" of his own machine, so that he should not be alarmed if he experiences sudden gusts and bumps during these displays, for they are only caused by the air disturbances created by his own machine and the small amount of air space in which he is manœuvring.

Trick fliers at exhibitions have sometimes done such things as under loops and flying upside down; but the machines are generally specially constructed and strengthened for such work. Provided that they allow themselves enough height, there is practically no positional manœuvre which it is not possible to perform on a machine, or from which the machine will not right itself—generally more or less automatically.

Opinions are divided as to the advisability of allowing pupils to attempt the easier of these air tricks; but, provided they are high enough in the air, such practice not only gives them added confidence in themselves and in their machines, but also undoubtedly makes them better pilots. A pupil should obtain his instructor's permission before he attempts even the simplest of these manœuvres.

Safety in Height

For war purposes and aerial fighting, the man who can manœuvre his machine the quickest obviously stands the best chance of downing his adversaries. On the other hand, no stunting near the ground or over bad country at low altitudes should be attempted, for, sooner or later, the pilot will miscalculate his speed or distance, or else his engine may fail when he is at a critical angle, and he is unable to right himself before he crashes into the earth. More than one pilot has been severely injured or even killed through taking unnecessary risks such as these, or in his desire to prove himself a finer pilot than a rival. If any useful purpose were to be served by such displays, then by all means let them be attempted, even though the risk be great; but when it is only a matter of personal rivalry, or display before an admiring crowd, then such tricks are sheer foolhardiness.

CHAPTER XI

Night Flying

THE question of night flying may now be discussed, the pupil having already been taken through the various stages of school flying until he may be considered to be tolerably expert. Incidentally, it may be remarked that even the finest pilots are constantly learning and adding fresh knowledge to their minds gathered from the hard school of practical experience.

Night flying has often been regarded as a dangerous pursuit, probably owing to the number of accidents that occurred when it was first attempted on a large scale under unsuitable conditions. The lessons taught by these accidents have brought about a greatly-improved state of affairs, and there is now no particular danger in practising night flying. This is a very different matter to undertaking a night flight under bad weather conditions against hostile aircraft.

Aerodrome Lighting

The first thing for a pupil to remember is not to attempt to fly at night until he can land a machine by looking at the horizon, i.e., by looking ahead for a distance of at least 100 yards. At any time it is a mistake to watch the ground close under the machine, where it seems to be travelling very fast, and where an idea of the height that the machine is above the earth is difficult to estimate. In night flying this mistake must be absolutely avoided. As petrol flares are usually arranged on the night landing ground in the form of the letter "L," one can take the tail of the letter as the horizon in this case, and the long side as show-

200

ing the extent of the landing ground and the direction in which to land.

The next point concerns the machine. Practically all night flying done in England is done on stable machines, i.e., a machine that will practically fly, land and bank itself automatically. Obviously, the strain of flying at night is much reduced by the use of such machines, which are also capable of automatically righting themselves if they should get into a dangerous position.

It is a good plan for the prospective night pilot to practise flying and landing at dusk. He can then carry on with his flying later and later, until, when he gets a fine moonlight night, he will notice little or no difference between his flying conditions then and at dusk; indeed, many pilots think that dusk flying is more difficult owing to the rapidly-changing light. It is also obvious that he must know by heart his own aerodrome and any traps here may be in getting off or landing. He can familiarise himself with other aerodromes in the neighbourhood.

Daylight Practising

It goes almost without saying that a night pilot must become familiar with the handling of his machine by day before attempting to fly at night. This remark applies almost to any piece of mechanism, for the more the man and machine are one or in sympathy, the better will be the results obtained from them. Therefore, the night flyer should practise all day what he intends to do at night on the identical machine he will use. The machine should be rigged in such a manner that it will climb slightly if the control lever is let go. The object of this is that the machine will fly at a more or less uniform and medium speed. It should be adjusted to fly level with the throttle half shut. In practising for night flying by day it is a good plan for the pilot to watch his instruments more carefully than he would for daylight flying. He can practise watching his sideslip and air speed indicator and correct the machine by these. He should also watch his altimeter creep up or down according to the variation in the height of the machine, and note how it lags when the machine descends rapidly.

Getting Off at Night

In getting off at night it is very important for him to remain stationary for at least one minute at the end of the L lights, facing the short bottom end of the L, which will indicate the length of his run and impress on his mind the lie of the landing ground which he will have to use when he returns. He should count the number of lights and calculate the distance allowed him. He can also take note of any particular obstacles near at hand which he will have to avoid either in landing or getting off. In getting off he must be very careful not to swing. If possible, he should fix his eye on some distant light to steer for, and he should never attempt to turn under a height of 500 ft. or 1000 ft. He should not take the machine off the ground, but, instead, he should allow it to fly itself off. This is very important indeed. He can practise this by daylight and will find it quite a simple operation. The proper time to leave the ground is the moment when it is difficult to hold the machine down any longer.

Keeping the Flares in View

In the air he should make certain that he keeps the flares if he wishes to keep them in view continuously, not fly over them, but rather to one side. Up to a height of 6000 ft. he should pay particular attention to this, but must remember to keep right away from the flares if he wishes to keep them in view continuously. At the same time, he must keep within gliding distance of the aerodrome, so that, if his engine fails or if unfavourable weather conditions begin to appear, he can be certain of his landing ground. If there is any suspicion of a ground mist he should come down at once. If more then one machine is practising at the same time, he should arrange beforehand some kind of signal or lighting whereby each pilot can recognise the other's machine. Alternatively, each pilot can arrange to fly in a certain direction so as not to interfere with the other. Very's light may be fired from time to time if there is nothing better at hand. All night pilot pupils should avoid clouds so far as is possible, for they will obscure the aerodrome flares and the pilot may lose himself, or find himself in an uncomfortable position through losing his horizon. If the clouds are high up this does not matter so much, as he can easily come down and find out where he is.

Night Landing

Landing at night is the most difficult part of night flying. The pupil should never attempt any kind of landing except the straight glide. "S" turns at night are dangerous, and spirals are quite unnecessary. He should start his straight glide into the aerodrome from a couple of miles away, with his engine on slightly, very much in the same way as he would attempt to land in a small field in daylight. He will thus half fly, half glide towards the aerodrome. When he thinks he is within 30 ft. or 40 ft. of the ground on the near side of the aerodrome, he must shut off his engine completely, but he must not put his nose down. Instead, he should keep the control lever in exactly the same position as with the engine doing about half its normal number of revolutions, as was the case on the descent. He should use the lights at the short arm of the L at the end of the aerodrome as the best means of judging the height, and, at the same time, he must keep well away from the lights and the long arm of the L down which he is landing, and which gives him his distance. If he does not keep away from them he will not be able to see them. He should start flattening out early, but should do so gently and gradually. If he does not touch the ground when he thinks he ought to, he should pull the control-lever right back, but not so as to balloon, and await the bump.

If he does not hit the ground by the time that the third flare is passed, he must switch on and open up his engine again to make another circuit. It is advisable to land short rather than over, the ideal spot for landing being about the second flare. A landing should never be made with the engine on, as this increases the speed of the machine. People standing near the flares will give him some

idea of his height; but this is not so accurate as that given
by the landing flares at the end of the ground.

Night-flying Equipment

A few words on the equipment of the night-flying ma
chine may be useful. The electric-lighting outfit for the
instruments should be duplicated throughtout, two lights be
ing fitted to each instrument and the switch duplicated
The instruments themselves should have the figures painted
with luminous paint.

An electric launching tube complete with battery and two
parachute flares should be taken in case of a forced land
ing. The tube may be conveniently fitted between the pilot's
legs, and a stick should also be carried to poke the parachute
flare through the tube in the event of it sticking. If he has
to make a forced landing, he must remember that the wind
will blow the parachute flare in a certain direction; there
fore, he must let off the flare at a height of 1000 ft. or so
outdistance it, and then turn and land into the wind against
it.

Holt's wing-tip flares are often fitted, but must be placed
so that there is no chance of them setting the wing on fire.
Sometimes the wing tips are made of aluminium to prevent
this. Flares burn for 60 seconds, and can be switched on at
a height of 1000 ft. Some machines are arranged with a
series of electric lights and powerful reflectors along and
under the leading edge of the wings, the pilot being able to
control their exact position by wires, so that he can turn
them down when he is near the earth and see what the
ground is like. On some pusher machines a system of il-
lumination is fitted on the under-carriage.

It is important that the propeller should be painted black
so that the light is not reflected on it, and does not dazzle
the pilot's eyes when landing. Any bright parts of a machine should also be painted black for the same reason.
Wings, fuselage and tail are often painted black for this reason, and also to prevent the machine being seen by enemy
aircraft.

Setting Out an Aerodrome

The setting out of an aerodrome for night landing is a very important matter. The petrol flares used to illuminate the ground are placed in the form of the letter "L," generally with the long arm lying into the wind. In certain circumstances it may be placed slightly across wind, as, for instance, if there is a better entrance to the aerodrome in

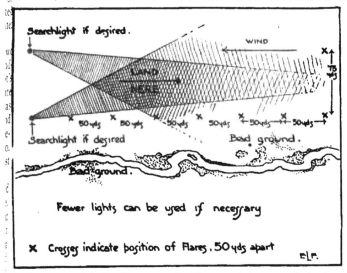

Setting out a night landing ground. The petrol flares are in buckets and arranged at intervals of 50 or 100 yards. The tail of the L-shape indicates the limit of good landing ground, and also the direction from which the wind is blowing.

that direction and the wind is only light. The short arm always gives the pilot the far end limit and the best area in which to land, the length of this between the flares being about 75 yds. The distance between the flares on the long arm is 50 yds., the number of flares used depending on the length of the aerodrome. It does not matter whether the long arm is placed to the right or the left of the short arm, but it is generally arranged to act as a guard against such obstacles as trees, houses or sheds when they border the aerodrome. Different aerodromes can be recognised at night by the different distances adopted between the flares. At the two near corners of the aerodrome, i.e., where the pilot is to come in, two searchlights are sometimes arranged with their beams converging at a point about one-third up the line of flares, which is the proper place for a pilot to land. It is a good plan to place a high pole, with a row of electric lights down it, at the near side of the aerodrome (but not so as to interfere with the pilot), to give him an idea of height, while a bonfire 500 yds. or 600 yds. away, to illuminate the country and obstacles in front of him when finishing his glide, is also of assistance. All lights should be shaded so that their glare should not dazzle the pilot when landing. The illustration (page 205) shows how to set out an aerodrome for night flying.

Night Flying Tests

A prospective night pilot is seldom given any dual control for night flying. His instructor can tell by his daylight flying if he is likely to make a successful night pilot, although it does not follow that because a pupil is only a fair daylight pilot he will be no use for night flying. He is generally considered fit for night flying, under moderate conditions, if he can make six or ten successful landings and attain a height of 6000 ft. in the dark. He must familiarise himself by daylight with the aerodromes equipped for night flying where he may have to land. He must possess a good knowledge of the country and be familiar with those parts of it which are open, and those which are wooded or otherwise dangerous. He can acquire this knowledge by moonlight nights as well as by day.

If an engine fails on a dark night, some people consider that the safest place to come down is on water, where the machine can be pancaked.

When flying at heights of, say, 2000 ft., navigation lights of red on the left wing and green on the right, with a white one on the tail, can be used so as to avoid other machines.

CHAPTER XII

The Growth of Confidence.—A Word Both to Instructors and Pupils

ALTHOUGH an instructor cannot make a good pilot out of an unsuitable pupil (as a pilot is born and not made), he can prevent a moderate pupil from becoming a moderate pilot by unwise actions. The art of instructing lies in getting the best out of the available material at one's disposal. For instance, out of 20 pupils, two may turn out first-rate pilots, four distinctly above the average, six may give up aviation before they have completed their training, and the remainder will be in the average run of pilots.

Instilling Confidence

A good instructor will be able to decide in the early stage of tuition which of the 20 pupils are likely to be useless as aviators, and the sooner that they are got rid of the better for both instructor and pupil. If they hang on, disliking flying all the time, yet being afraid to acknowledge it, a tremendous amount of time and money is wasted upon their tuition. Not only that, but it is more than likely that they will inspire similar feelings in some of the other pupils, with the result that others, who, had they not been subjected to this baneful influence, would have made aviators of moderate skill, may also become nervous and lose their original keenness.

The growth of confidence in a pupil is one of the most important points which the instructor must study. He must try and inspire the pupil with confidence in himself as instructor and in his judgment of the pupil's own capabilities; while later on, when the pupil has flown solo, he must study the pupil's own confidence in himself. For this reason the pupil's first solo is of the most vital importance, and he should not be sent up unless he feels confidence in himself. Let the instructor remember that an hour's extra dual-control work—which may make all the difference between a crashed machine, an accident to the pupil, with the possibility of the latter resigning afterwards, and a first-rate pilot—is always worth the time and trouble expended.

Steady Progress

As the pupil advances in his course he can be sent up for longer and higher flights, the call upon his capabilities being gradually yet surely increased. Every time the pupil comes down after successfully doing something in the air that he has never done before, his confidence in his own abilities will have been increased. It may have been that he has done a spiral for the first time, or gone up in bumpy weather, or climbed a few thousand feet higher than he had ever done before; but in every case the growth of confidence will be there, and it should be the instructor's work to study it and to make the pupil make the best advantage of it in pushing his training ahead.

On the other hand, a pupil should remember that a good instructor will probably know much more about his capabilities as an aviator, at any rate in his early stages, than he does himself. The instructor can tell after a few days which pupils are keen and which are not; and here it may be repeated that a pupil who, after a few hours dual flying, decides that he does not like the work, had better say so and resign at once, rather than continue to waste his own and instructor's time pretending that he does. There is no disgrace in owning up to a feeling of nervousness in the air, and if more pupils did this as soon as they felt certain they would never make aviators, it would save a great deal of time and trouble. A first-rate instructor will generally turn out first-rate pupils, and vice versa; so that a nervous pupil may possibly be made much worse by a nervous instructor, whereas if he were sent to an experienced pilot, who would allow him to do anything he liked in the air, he might overcome his nervousness.

After a Crash

Sometimes a pupil, in the early stages of solo work, has an accident and loses confidence in himself. It may pay to send him away for a few days' rest and change of scene, so that he cannot brood over the crash; but if he is of a certain temperament, even this will not cure him, and he may have to resign altogether.

Another type of pupil would not be affected at all in the same circumstances, but would want to go up again at once and make good the fault he had just committed. Probably the age of the pupil is largely responsible for this difference in temperament; a young and not yet fully-developed pupil might be permanently frightened by a crash which an older and more experienced man would think nothing of.

The More Practice the Better

It is a very curious fact about aviation that the more one flies the less one minds it, and the less one flies the less one feels inclined to, so that it rests largely upon the pupil himself to go ahead. If he can overcome his apprehensions and master them, he will feel extra confidence, and the next time he will not find quite so much difficulty in forcing himself into the air. Eventually, if he repeats the process a sufficient number of times, he will think as little of making a flight as getting into a train or taking a taxi.

Recreation and, if possible, a change of scene are useful in getting the best out of pupils. It is a bad thing when an accident has taken place for pupils to be able to congregate and discuss all the details among themselves. It would be far better if they were able to go away and play a game of football or cricket, and so forget the accident, otherwise the less keen and stout-hearted among them will begin to brood, and, possibly, in the end, decide that they had better not continue with their course.

Some pupils become nervous after they have done several hours solo work in the air. This may be due to something happening to them in the air which they do not understand, or else it may be brought on by their being pushed forward too rapidly. A rest, in the latter case, may be advisable, when they may return to their work with all their original keenness. It all comes back to the original question, that it rests with the instructor as to what course he should advise with reference to any given pupil. A course of action with one may prove fatal to another, hence the need for a careful study of the individual pupil's character and capabilities.

Dual Control in Rough Weather

Flying in bumps is a difficulty with some beginners. Often it is a good plan for the instructor to take up a pupil in a dual-control machine in bumpy weather and to allow him to fly it so as to discover for himself that there is no particular danger in bumps, although the feeling of security and conrol may not be so great as it is in calm weather.

A point of great importance is to get the pupil over his first two or three hours solo work as quickly and as safely as possible. If his first few landings or get-offs are not sufficiently good, the instructor should explain to him his faults, and, if necessary, give him more dual-control. If bad weather intervenes between the pupil's first few solos, more dual-control work should be given to allow the pupil to get his eye in again. For this reason, especially in winter time, the pupil should be pushed forward over his earliest stages of flying, so that he can gain the necessary confidence in himself as soon as possible, otherwise his course of instruction under bad weather conditions may be prolonged for months instead of weeks.

Advance Dual Control

Even when a pupil can fly a machine solo, he may teach himself a number of bad faults, and for this reason an instructor would be well advised to take up the pupil from time to time and test his flying. Much more can be done nowadays by means of dual-control work than was thought possible in the early days. Pupils can be taught to do vertical turns and spirals, loops, spinning nose dives, cart-wheels, stalls, etc., but the principal point to remember is that the object of giving advanced dual-control instruction after the pupil can fly solo is to secure accuracy in flying, and to prevent the pupil from getting into careless ways of handling his machine in the air. Careless or bad flying may not matter much on a heavy and soggy elementary machine; but if the pupil allows himself to adopt the same sloppy methods on a machine of a more sensitive type, he will soon find himself in difficulties. Hence the importance of accurate flying even in the earliest stages. Another point in favour of giving a pupil advanced dual control is that this system has been found to increase his confidence in the machine in the air. If an instructor shows a pupil how to get out of practically any position, the pupil will be able to save himself if he ever finds himself in a similar position. Not only that, but the pupil will possess the necessary knowledge and confidence to be able to put the machine in those positions, besides being able to get it out of them. All this means that he will be a far better pilot.

When a Pupil Can Fly

A pupil should be able to fly an elementary school machine solo after, say, five hours dual-control instruction. He will then do 20 or 30 hours solo on various types of machines, with a little dual control in landings on the more elementary types. If he shows some aptitude for flying he will, at the end of his 15 or 20 hours solo, be put on to faster and faster machines of the scout type, in which case, of course, no dual control is given. The length of his course of instruction depends on the weather. In the summer time it may take him two or three months to qualify as a scout pilot; whilst in the winter time it may take him much longer. Naturally, the keen pupil will get ahead of those who are less enterprising, with the result that he will benefit accordingly. It rests with the pupil to improve his own flying after a certain point. The great thing is constant practice.

The Time Taken in Learning

A word to instructors on the quickest method of teaching pupils to fly may be of value at this point. If the pupil can be taught to fly in the air in, say, from 40 minutes to 50 minutes, the get-offs and landings may take him a further 3 hours or 4 hours. It is therefore advisable to get the pupil on to get-offs and landings as soon as possible. He will require more instruction in landing than in taking off. Therefore, to speed up instruction, the more landings the instructor can put in in a given time the sooner will the pupil become a pilot. With this in view, some instructors prefer to make large circuits of an aerodrome, landing, perhaps, four or five times per circuit and switching on again without stopping the machine. Others may do straights and land two or three times in each straight. A

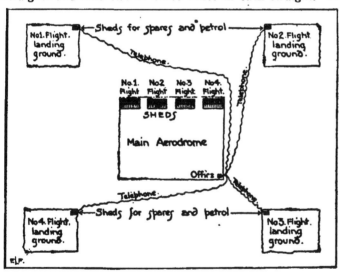

Plan of aerodrome and landing ground. Suggested arrangement of main aerodrome and subsidiary landing grounds at a distance of a few miles from headquarters. This arrangement should facilitate landing practice for a large number of pupils at one time.

third plan is to make small circuits, rising to a height of several hundred feet, and then land in the same place again and again.   The greatest number of landings can be got in in a given time by the first method, but this necessitates comparatively calm air, as otherwise the landings in the course of the circuit would have to be made across and down wind.   A point that should be noted is that a landing from a low height, say, 100 ft., does not teach the pupil much, as the landing he should practise should be prefaced by the ordinary glide of several hundred feet.   There is a good deal of difference between landing from a height of 100 ft. and from a height of 1000 ft., and if the pupil has only been taught the first, he will probably go hopelessly wrong when he attempts the second.

Thoroughness in Instruction

An instructor should always remember that a pupil must be taught every bit of flying that is necessary in handling a machine, i.e., starting up, taxying, taking off, climbing, flying level, turning to right and left, shutting off the engine, gliding and landing.   Furthermore, it may be necessary for an instructor to show a pupil what to do in emergencies, i.e., if he bounces through failing to flatten out, or if he should flatten out too soon or too much; then, when the pupil is faced with these situations, he will do the right and not the wrong thing.

When a very large number of pupils are being trained at the same school, it will prevent congestion of the air and aerodrome if a number of small landing grounds are selected at a radius of several miles from the air station.   It is far easier for 36 pupils to be given landing practice at six different fields than for the whole 36 to be taught on the same small patch of aerodrome.   The accompanying sketch gives an idea of how this suggestion might be carried out. (See opposite page.)

CHAPTER XIII

The Medical Aspect of Aviation

By

H. GRAEM ANDERSON, M.B., Ch.B., F.R.C.S.

Temp. Surgeon R.N., Attached R.N.A.S., Surgeon to St. Mark's
and the Belgrave Hospitals.

T O fly, an individual must be physically and temperamentally fit; especially is this so during instruction in the first 20 hours of solo flying. A pupil
should not conceal any disease, and should he feel physically
and temperamentally unfit, he should inform his instructor.

Many pupils remain silent and do not confess their disinclination to fly, fearing that their instructor or other pupils
will taunt them with having cold feet. "It is certain that
an aviator's disinclination to fly must have its basis upon
some temporary defect of body or mind, and, without being
unduly sensitive or timid, he should realise this and overcome the cause rather than tempt Providence by running
the danger of overtaxing his power." This quotation from
another writer on the subject applies most particularly to
pupils.

Flying is a question of an active, well-balanced, decisive
mind and a series of sound and quick reflex actions.

These reflexes are :—

1. Visual.   3. Tactile.
2. Auditory.   4. Muscular.
        5. Balancing.

The Visual Reflex

(1) The visual reflex consists of the impressions carried
through the eyes to the brain, and from there down to the
muscles of the hands and feet; e.g., the pupil sees a wing
tilted up to the right; this impression is transmitted through

214

the eyes to the brain, there recorded, and a decision made which is sent down to the muscles of 'the hands, and the control lever is pushed over to the right. The normal time for the visual reflex action to occur is 20-100ths of a second. This is the reflex action, the most important and most used by the aviator. It is essential that all parts of this reflex are in good working order. Thus it can be seen how important is good vision in flying. I consider that no person should take up aviation unless he has full normal vision in both eyes.

The Auditory Reflex

(2) The auditory reflex consists of the impressions carried through the organ of hearing—the ear—to the brain, and from there to the muscles of hands and feet; e.g., the sound of the engine missing is conveyed to the brain through the ear, recorded there, and a decision immediately made to look for a good landing place, land and repair the defect, is transmitted to the muscles of action of eyes, head, feet and hands. The normal time for the auditory reflex is 14-100ths of a second. This reflex is next in importance to the aviator, and it is essential that he should have good hearing. Wax is apt to accumulate in the ears and cause deafness. This deafness comes on gradually, but is usually made much worse after washing or swimming, when water gets in the ears and expands the wax. Lack of good hearing may lead to disastrous results. For example, when starting up the engine, should the pilot fail to hear—or misinterpret—the words "Contact!" or "Switch off!" when given by the mechanic who is swinging the propeller.

Tactile and Muscular Reflexes

(3) The tactile and (4) the muscular sense reflexes are the impressions conveyed through the nerve endings in the skin and muscles to the brain, and from there down to the muscles of the hands and feet. For example, the pilot feels a bump, which sensation is conveyed from the skin and muscles to the brain, recorded there, and the decision sent

down to the muscles of the hands to correct the effects of
the bump. The normal time for the tactile reflex is 14-
100ths of a second.

The Balancing Reflex

(5) The sense of balance or equilibration is received
from the semi-circular canals, three in number, on each
side, encased in bone, and situated behind the ears. These
canals contain fluid and fine nerve endings. When the pilot
and machine are off the level, either fore and aft or laterally,
these little canals send impressions to the brain, and from
there correcting impulses are sent down to the muscles of
the hands and feet.

All these reflexes are slowed down or disturbed if the
pilot is physically or temperamentally unfit from any
disease, worry, fatigue, or after excesses in alcohol, etc. A
slow reflex action, the delay of a second or part of a second
in correcting an error in the air or in landing, may mean
all the difference between a crash and safety. Thus, to preserve these reflexes, one sees how important it is to keep
the body and mind in good condition during the tuition period of flying.

Drinking and Smoking

Alcohol is better avoided altogether, and, similarly, excess in smoking, which may cause palpitation, faintness and
double vision. Most aviators smoke too much. Diet should
be generous and nourishing, as there is a good deal of nerve
strain and wear and tear of the nervous system during this
period. Flying when hungry is to be avoided, as faintness
may occur in the air. Proper sleep is most important, and 8
hours sound sleep in the 24 hours should be obtained. Well-
regulated physical training is of great value, and pupils
should be afforded every recreation of mind and body at an
air station.

With regard to the psychology of flying, or the study
of the sensations in the air, it has been found from an
analysis of 100 confessions of pupils after their first solo
flights that the mind is so occupied in paying attention
to flying, watching instruments, controls, etc., that fear has

rarely time to assert itself, at least, not enough to disturb their flying. Later on, a pupil is apt to become over-confident, and this must be guarded against. Coolness, level judgment and quick reflex actions are the secrets of success in flying. The fact that aeroplanes are now so improved and structurally strong that there is little or no danger of anything giving way in the air, should reassure pupils, who sometimes are distressed with this thought whilst in the air.

Pupils should always strap themselves in before getting off the ground, and they should become familiar with the fixing and unfixing of the belt. Belts should never be undone in the air. Unless belted, it is possible that in a bump, dive, or in faintness an aviator might fall forward on to the control lever with disastrous results. It is advisable for all pilots to carry a stout knife and a wire cutter in the outside pocket of the flying coat, so as to be able to cut themselves out of a crashed machine. (See page 183.)

What to Wear

Safety helmets should be worn by pupils, as, in the event of a crash, they prevent injuries to the head and ears caused by wires and broken struts. The helmet should fit properly and not be easily dislodged from the head so as to slip forward. The ears must be well protected in front so that there is no rush of air straight across the orifice of the ear. A good helmet has a little roll of leather in front of each ear to deflect the air stream, thus enabling the pilot to hear well and also to prevent any damage to the ear drum.

Goggles should always be fitted with non-splintering glass, such as Triplex. Plain glass goggles should never be used, as in the event of their breaking serious injury might be caused to the eyes.

Gloves of leather lined with lambswool are the best for warmth and do not diminish the touch. The gauntlets should not be too stiff or too wide. It is better to wear the gauntlet inside the coat cuff, the latter being drawn tight by a strap. Close and tight-fitting gloves should be avoided.

Soft leather field boots lined with lambswool and thick woollen socks keep the feet and legs warm. Rubber boots should not be used, as they do not give much warmth and are apt to slip on the rudder bar.

The body clothing for flying, of course, varies for winter and summer. No articles should fit too tightly. Woollen material is the best, as its interstices allow of a layer of warm air next to the skin. Leather outer garments are usually worn.

Most aviators fly with the mouth slightly open. Pupils should see that their teeth and gums are in a healthy state, otherwise any local disease therein is apt to be increased by the cold and rush of air.

Preventing Frostbite

Should a pupil have to do a height test in winter, a good method of preventing frostbite—or should the skin be sensitive—is to smear the face and hands with a thin layer of vaseline. This prevents heat loss from the skin.

Air Sickness

There are two forms of air sickness. One, which is akin to sea sickness, and due to the rolling and pitching of the aeroplane in bumpy weather, is very rare, but now and then one finds an individual liable to this ailment. Of course, a pilot, by doing steep spirals and switchbacks, may produce this form of sickness in a passenger. The other form of air sickness is due to height effects, and is better named "altitude sickness." It occurs at height of 10,000 ft. and over, and is caused by the rarefied atmosphere and lack of oxygen. After passing 6000 ft. the breathing and pulse rate become quicker. Over 10,000 ft. the cold is extreme; slight buzzing in the ears, difficulty in hearing, headache, fatigue and torpor may occur. On descending rapidly, the deafness and buzzing in the ears become more acute, and severe earache may be felt. The headache continues for some time after landing and sleepiness is very marked. To guard against these effects when going to great heights, especially for any length of time, oxygen should be carried and inhaled slowly as a preventive.

FLYING INSTRUCTION NOTES IN BRIEF

General and Condensed Hints in Definite Stages for the Guidance of Pupils

STAGE 1.—BEFORE INSTRUCTION—GENERAL

1.—Find out why a machine flies and learn to understand what happens when you get off, fly level, dive, turn steeply or gently, and land.

2.—Study the control levers and their effect when moved on the controlling surfaces (rudder, elevator and ailerons) of the machine you are going to fly.

3.—Find out how the switch, petrol and oil taps, throttle and fine adjustment work.

4.—Find out how the instruments work and what they are for; air speed indicator, height recorder, engine revolution counter, compass and sideslip indicator.

5.—Learn all about the details of construction of the machine and engine.

6.—Practise swinging the propeller.

7.—Practise sitting in the machine and working the control levers as if you were flying.

8.—Find out the correct number of revolutions at which to run the engine, and the climbing, flying level, and gliding speeds of the machine.

9.—If there are any points that puzzle you, ask your instructor to explain them.

10.—When an engine or machine gives trouble, find out the cause and remedy.

11.—Obtain a map of the country round the aerodrome and study it, so that you can pick up your bearings in the air.

STAGE 2.—EARLY INSTRUCTION

1.—Never keep your instructor waiting. Be ready with flying gear, speaking tubes, gloves, cap, goggles, etc., to take your place in the machine.

2.—Always be punctual at the shed.

3.—Before going up, fasten your safety belt.

4.—Get a clear understanding with your instructor as to what signs or signals he will make when you are to take control, let go, fly straight, or turn.

5.—When given control, do not hold too tightly, but do not be frightened of working the controls when necessary. You can fly the machine with one hand and two feet.

6.—Try and get the "feel" of the machine as soon as possible.

7.—Always look out in front of you on the horizon. It gives you the best guide as to when you are flying horizontally or longitudinally level.

8.—If you don't understand, always ask your instructor to explain anything he has told you, or any point that has occurred to you during your flight.

9.—To fly horizontally level, keep the nose of the machine on the horizon. To climb, keep the nose above the horizon, and to descend put the nose below the horizon.

10.—To fly laterally or horizontally level, keep the underside of the top plane parallel with the horizon.

STAGE 3.—FLYING LEVEL AND TURNS

1.—To fly straight, fix your eye on some landmark on the ground ahead of you, such as a village or wood, or a cloud, and steer the machine towards it by means of the rudder.

2.—The position of the control levers is only relative. If you find that the machine is flying to the left in still air, with the rudder central, you must give a little more right rudder to fly straight.

3.—Learn to fly by sight and feel. Only use your instruments to verify occasionally the accuracy of your flying.

4.—Don't be alarmed at "bumps," but correct them quickly and firmly but not jerkily.

5.—In turning, keep the nose of the machine on the horizon. Then apply bank and rudder gently and gradually in the direction you wish to turn. If your bank is too steep ease off the control lever. If your bank is too small, put on a little more by moving the control lever still further over in the desired direction.

6.—When straightening up after turning, move the control lever slightly in advance of the rudder, and rather past the central in the opposite direction to begin with. It can then be moved back to central when the rudder is central and the machine horizontally level.

7.—Learn to make gentle and gradual turns before attempting quick ones, which need more bank and greater accuracy.

8.—If you are turning to the right and notice wind striking your left cheek, you are side-slipping outwards, so give a little more bank or take off some rudder. If the wind strikes your right cheek on a right-hand turn, you are side-slipping inwards. Remedy: more rudder or less bank.

9.—In making steep turns over 45 degrees, first bank and rudder in the desired direction. Trim the nose of the machine to the horizon by easing off bottom rudder, and make the machine turn by pulling the stick back towards the opposite elbow.

10.—As the machine will be on its side relative to the horizon, the rudder acts as elevator and the elevator as rudder.

11.—To come out of a steep turn, move the control lever still further over in the opposite direction and then forward and back to central position in a semi-circular movement. If you centre the control too soon, the machine will not be flying horizontally level. If you give opposite rudder too early, the nose will go up and you will side-slip inwards.

12.—In climbing and gliding turns, the nose of the machine must be kept above and below the horizon respectively, according to the correct climbing and gliding angle. The same rules re making and coming out of these turns apply as in flying level turns.

STAGE 4.—GETTING OFF

1.—See that there are no loose articles in the passenger's seat that can foul the controls.

2.—See that the chocks are under the wheels.

3.—Fasten your belt and set the height indicator to "o." Hold the control lever back.

4.—Turn on petrol and see that the switch is off. Pump up pressure if necessary.

5.—Open throttle or fine adjustment and close air to suck in.

6.—Close throttle or fine adjustment, let go air lever, and give contact when the mechanic at the propeller is ready.

7.—Note the formula adopted for starting up the propeller, i.e., "contact," "switch off," as said by the mechanic and repeated loudly by the pilot.

8.—Don't run the engine more than is necessary. Warm up from cold gradually and throttle down again as soon as maximum revolutions are obtained.

9.—Wave your hand over your head to mechanics to withdraw chocks, and see that they have let go.

10.—Taxy out slowly with stick rather back of central position.

11.—Turn by using the rudder with a little extra engine.

12.—Get off into the wind.

13.—Get under way gradually, i.e., open up the engine steadily and not with a jerk. Some engines give their correct revolutions when the throttle is partially open.

14.—Push the stick forward as the machine gathers speed, and then ease it back when the tail is up and you think you have attained sufficient flying speed on the ground. It is safer to get off too late than too early.

15.—When it is difficult to hold the machine on the ground, the control lever can be eased back, and the machine will take the air. Don't pull the lever back with a jerk.

16.—As soon as the machine is in the air, ease the control lever forward, so as to allow the machine to gather more speed. Then climb steadily to the required height, say, 1000 ft.

17.—If the machine tends to swing sideways when getting up speed on the ground, rudder in the opposite direction and counteract the swing before it develops.

STAGE 5.—LANDING

1.—Close the throttle fully and put the nose of the machine down to the correct gliding angle, i.e., with the nose the correct amount below the horizon. The speed should be slightly under flying level speed.

2.—Make a straight glide.

3.—When approaching the earth, watch the ground 30 yards in front of the machine. *This is very important.*

4.—Don't be disheartened if you fail to make a good landing to begin with. You need more practice.

5.—The ideal flatten out starts with the control lever in the gliding position about 30 feet from the ground, and ends with the control lever right back in your chest as the machine touches the ground.

6. Try and keep the wheels of the machine *off* the ground as long as possible.

7.—The machine must not lose its flying speed until it is a few inches from the ground.

8.—If it loses its flying speed too early, it will pancake, that is, you have flattened out too soon, and you will flop instead of gliding to earth.

9.—If you try and lose your flying speed too late, you will fly into the ground and bounce.

10.—A good landing is all a matter of an accurate eye and a fine touch. No sudden or jerky movement is required.

11.—If you bounce badly, use your engine, make another circuit and try again.

12.—To begin with, the ground will seem to be coming up very fast towards you, but you will soon get used to it.

13.—In landing or getting off cross wind, keep the wing nearest the wind down, so as to eliminate a crabbing motion over the ground by side-slipping into the wind.

Stage 6.—Taxying

1.—Taxy slowly, with the control lever well back, to keep the tail on the ground.

2.—Turn the machine by means of the rudder.

3.—Look where you are going, and go dead slow over bad ground or ruts.

4.—Use your motor well throttled down, and don't open up or throttle down in jerks. Don't "blip" the engine unless you have been taught.

5.—When machines are getting off or landing near you, keep stationary so as not to baulk them.

6.—Don't turn quickly on the ground, as this wears out the tail skids and twists the tail.

7.—In taxying down wind, keep the control lever forward, so that the wind on the elevator keeps the tail down.

8.—In taxying in a strong wind, tell off two mechanics to hold on to the wing tips, so as to prevent the machine lifting.

Stage 7.—The First Solo

1.—Don't be in two much of a hurry. Fasten your safety belt and see that your crash helmet is secure.

2.—Clearly understand your instructor's final advice as to where you are to get off and land, how long you are to stay up, and at what height you are to fly.

3.—See that your motor is doing its correct number of revolutions on the ground, and when you have attained the desired height throttle down to fly level. Set the height indicator to "o."

4.—Give the proper signal to the mechanics to withdraw the chocks by waving the hand over the head, and wait a moment or two for them to get clear.

5.—See that there are no machines getting off or landing which may baulk you in starting.

6.—Remember the instructions on getting off. (Stage 4.)

7.—Don't go out of gliding distance of the aerodrome, but make full circuits, and don't cut across the aerodrome or go the wrong way round.

8.—Look out for other machines in the air.

9.—You will find that the machine will climb much more quickly solo than dual, so don't get too high before you realise it.

10.—If the weather is doubtful, look out for rain or thunderstorms or banks of mist, and come down at once if they approach the aerodrome.

11.—If you are lost, come down at once, but fly low, say, 500 feet, round the field you have selected to land in before doing so, in order to see that it is suitable for the purpose. Then find out where you are and return home. If you have damaged the machine, telephone immediately full particulars to the aerodrome.

12.—When you decide to land, make a straight glide into the wind.

13.—Remember the instructions on landing. (Stage 5.)

14.—If there are machines on the ground when you want to land, make another circuit till the course is clear for you.

15.—You should know the whereabouts of bad ground on the aerodrome, and must avoid it.

16.—If you bounce badly, make another circuit and try again.

17.—Don't point the machine at sheds or woods when landing, as you may overshoot and run into them.

18.—If you see that you will overshoot the mark, make another circuit and try again.

19.—Don't turn under 500 ft., and remember the instructions on turning. (Stage 3.)

20.—Don't get off facing woods or sheds unless you have plenty of room to gain height before crossing them.

21.—Report to your instructor after the flight, and listen carefully to criticisms, so that you can learn how to avoid such mistakes another time.

22.—Don't get over-confident. Take just as much care with your solo flights as you did with your first.

23.—Remember any special ground signs that are used to call all machines in from the air.

24.—Always report any defect in the machine to your instructor.

STAGE 8.—CROSS-COUNTRY

1.—Work out your course accurately and carefully.

2.—Spend half an hour studying the map and the course you have been set.

3.—See that the engine, machine and instruments are in good working condition.

4.—Take the necessary tools, spare parts and money with you. An identification card is advisable.

5.—Find out the distance of the flight, and take your time of departure so that you have a check on your journey.

6.—See that the petrol and oil tanks are full.

7.—Take nothing for granted, but go over everything yourself.

8.—Find your way by using the compass and map in conjunction. Avoid clouds.

9.—As soon as you are lost, circle round some well-defined landmark on the ground about where you should be judging by your time out and speed, and try and recognise it on the map. If you cannot, come down. (See instructions on Stage 7.)

10.—Always note the direction of the wind by smoke, or by what direction it was blowing when you last noticed it or at the start. You can also tell the direction of the wind by watching the drift of your machine.

11.—If your engine fails suddenly, switch off, turn the petrol off, and put the nose down.

12.—Select a field (grass for preference) free from obstacles, trees, houses, etc., and with length in the direction of the wind. A group of fields is preferable, as if you miss one you can land in the next.

13.—"S" turns provide the best method of making certain of hitting off the desired field.

14.—But don't turn near the ground.

15.—Aim at the near side of the field so as to give yourself the greatest run for landing.

16.—Don't come down too fast, as if you hit anything after landing or while gliding near the ground your speed will cause a worse crash.

17.—Don't land near railways or main roads for fear of telegraph wires, which are hard to see.

18.—Don't spiral down, as you may be facing the wrong way when you want to land.

19.—If you crash, telephone the aerodrome, stating the *full* extent of the damage.

20.—If you are able to continue the journey, take the fullest possible run in which to get off, and survey the ground ahead of you before starting.

21.—You must pancake in soft plough, standing crops or woods, with your stick right back and tail down.

22.—If after landing you see you are going to run into something, rudder away from it. You may remove the under-carriage, but this is better than smashing the whole machine.

23.—If you are up late, come down before it is dark.

24.—Peg the machine down for the night under the lee of trees, a hedge or house.

25.—Cover the engine, propeller and pilot's seat with a tarpaulin or sacking.

26.—If the motor goes wrong, try and locate the fault, and if possible repair it yourself.

27.—If your motor shows signs of gradually dying away, remember it is far easier to land with some engine than none, so come down and find the trouble. It may be a choked petrol supply or a faulty plug or two, or even loss of pressure in the petrol tank.

28.—Come down if there are signs of fog, heavy thunderstorms or snow.

29.—Practise map-reading before going on a cross-country flight. This can be done on solo flights round the aerodrome.

30.—In getting out of a small field it is best to "hoik" at the last moment and jump the hedge.

31.—Never stunt near the ground.

Stage 9.—Advanced Flying

1.—Every time you go up solo try and do something you have never done before, i.e., go up higher, or for a longer flight, or in worse weather, or practise aerobatics.

2.—First perfect yourself in landing on a mark with engine cut off.

3.—For this, "S" turns, spirals and vertical banks with and without engine must be practised.

4.—Practise climbing turns and stalling turns. The latter are performed by pulling the machine up on its nose and then banking and ruddering in the desired direction. The engine can be cut off during the manœuvre. Modifications of this manœuvre with the engine on result in cartwheels and Immelman turns in which the machine does an about turn.

5.—In all aerobatics avoid too sudden changes of direction of the machine, i.e., pull the machine out of a dive gently and gradually.

6.—To zoom, fly level at full speed, and then pull the stick back. At the top of the zoom ease the stick forward.

7.—A hoiked turn is a curving zoom, but you should know your machine, and practise this at 1000 ft.

8.—To loop, put the nose down to 85 m.p.h., and pull the stick straight back into your chest. When over, cut out the

engine, and pull the machine out of the dive gently. The
latter part of the pull back is quicker than the first in
starting the loop.

9.—To spin, cut out the engine, and stall the machine
level with the nose on the horizon. Kick on rudder, and
pull the stick right back into the chest. To come out, centre
the rudder gently, and ease the stick forward.

10.—To roll, get up speed either with or without engine,
kick on rudder, and pull the stick right back.

11.—To turn very quickly, switch off momentarily, kick
on rudder without any bank, which will pull the machine up
and twist its nose round in the desired direction. She
will get into a spin if not righted immediately. Then switch
on.

12.—To sideslip or partially glide sideways, switch off,
put on bank in one direction and rudder in the other. The
machine will glide sideways in the direction in which the
stick is held.

13.—The falling leaf descent is done with the engine off,
and the stick held back in the pilot's chest. After she stalls,
she can be allowed to gain speed, and then stalled again. If
the engine stops, dive to restart it.

14.—Diving is best done with the engine off. Put the
stick forward, and hold it there while the speed increases.
Pull out of a vertical dive very gradually.

15.—Try all advanced flying at a height of 2000 ft. or
3000 ft.

16.—Practise sham flights and aerobatics with other
pupils if your instructor thinks you are good enough. Practise approaching machines unseen, and always take note
of the machines in the air about you.

GLOSSARY OF TERMS COMMONLY USED IN AVIATION.

Aerobatics
> Aerial manœuvre stunts.

Aeroplane
> A heavier-than-air flying machine, supported by the action of air on fixed planes.

Ailerons
> Hinged flaps let into the extremities of the main planes and operated by the control lever, to bank the machine and also to maintain its lateral level.

Ailerons, Balanced
> By connecting the ailerons of each wing, so that when one is pulled down and the other is pulled up the surfaces are made to balance.

Airbrake
> A flap that can be let down so as to increase the resistance of the machine to the air.

Air Pocket
> See "Pocket."

Air Speed
> The speed of the machine through the air.

Air Speed Indicator
> An instrument for registering the speed of the machine through the air.

Altimeter
> An instrument for indicating the height of the machine from the ground where it started from.

Angle of Incidence
> See "Incidence."

Aspect Ratio
> The proportion of span to chord of a plane.

229

BACK
> A change of wind in an anti-clockwise direction, i.e., from E. to N.

BACKWASH
> The disturbed air in the wake of a machine in flight.

BALLOON, To
> The upward glide of a machine near the ground caused by the pilot descending too fast and pulling the control lever back too much or too quickly.

BANK, To
> To raise one wing for the purpose of turning.

BAY
> The space enclosed by two struts and their upper and lower adjoining surfaces.

BELT
> The safety strap which secures the pilot to his seat.

BESSONEAU
> A tent for storing aeroplanes which can be erected and dismantled in a few days.

BIPLANE
> An aeroplane with two pairs of wings set one above the other.

BLIMP
> Slang term referring to small airships.

BLIP, To
> To switch on and off rapidly.

BODY
> That part of a machine which accommodates the engine, pilot, passenger and probably the petrol and oil tanks.

BOOM
> See "Tail Boom."

BOSS OF A PROPELLER
> The centre portion by which it is attached to the engine.

BOUNCE

The upward and forward movement of a machine which has struck the ground without flattening out sufficiently.

BREVET

A certificate showing that a pilot has passed certain elementary flying tests and may be considered a qualified pilot.

BUMPS

Disturbances or roughness in the air due either to changes of temperature, clouds or wind.

CABANE

The projecting arrangement of struts above the pilot's head on a monoplane to which the anti-lift wires are attached.

CABRÉ

Tail down.

CAMBER

The maximum depth of curvature of the upper and lower surfaces of a wing.

CARTWHEEL

A particular type of aerial manœuvre.

CELL

The whole of the lower surface of a plane and the whole of the top surface of the plane above it, with the struts and wires holding them together.

CELLULE

The box-like rectangular compartments in a biplane formed by the upper and lower planes and the inter-plane struts.

CENTRE OF GRAVITY

Centre of weight.

CENTRE OF PRESSURE

A line running from wing tip to wing tip, through which all the air forces on the wing may be said to act.

CENTRE SECTION

The centre cellule of a biplane where this cellule is made detachable from the wings.

CHOCKS
> Wooden blocks placed in front of the wheels of a machine to prevent it moving when the engine is started.

CHORD OF A WING
> The distance between the leading and trailing edge of a wing.

COCKPIT
> The pilot's seat.

COLD FEET
> A complaint, otherwise known as aerosthenia or nervousness of going into the air.

COMBUSTION CHAMBER
> The space between the top of the piston and the cylinder where the explosion of the mixture takes place.

COMPRESSION
> The upward stroke of the piston which compresses the mixture in the combustion chamber.

CONK
> The engine is said to "conk" when it fails.

CONTACT
> Word used to denote that the switch is on.

CONTROL LEVER
> Generally known as the "joy stick" or stick. A vertical lever controlling the fore-and-aft and lateral movements of the machine.

CONTROL WIRES
> Wires connecting the rudder bar and control lever with their respective controlling surfaces.

COWL
> A sheet-metal cover generally fitted over or round the engine.

CRASH HELMET
> A specially-made flying helmet designed to save the pilot's head in case of a crash.

CRASH, To
> To smash the machine.

DIHEDRAL ANGLE

A machine is said to possess a dihedral angle when the wings rise upward from the centre of the machine.

DIVE

To descend steeply.

DOPE

A preparation used to paint the wings in order to render them taut and weatherproof.

DOPE CAN

A metal syringe containing petrol for priming the engine.

DRIFT

The crabwise motion of a machine over the ground due to a side wind; also used to denote head resistance.

DUAL CONTROL

A system of levers and controls for the engine and machine, so that either the pilot or passenger can operate them.

ELEVATOR

A hinged controlling surface, or flap, operated by the fore-and-aft movement of the control lever. Always set parallel with the wings of the machine and generally behind them. Used to control the up-and-down motion of the machine and in steep banks to make the machine turn.

EMPENAGE

The tail unit of a machine, consisting of rudder, elevator and fixed tail plane.

ENGINE BEARER

The metal framework or tubing to which the engine is fixed.

EXHAUST

The upward stroke of the piston which drives the burnt and exhaust gas out of the combustion chamber.

EXPLOSION

The power stroke of the engine.

EXTENSIONS
Additional lifting surfaces added to the top planes.

FACTOR OF SAFETY
Obtained by dividing the stress at which a body will collapse by the maximum stress it will be called upon to bear.

FIN
A fixed vertical plane generally fitted in front of the rudder to increase the stability of the machine.

FLARES, GROUND
Waste soaked in petrol, or petrol in buckets, set on fire and used as a landing light for night flying.

FLARES, PARACHUTE
Magnesium light electrically fired and attached to a parachute, which is released near the ground to facilitate landing at night.

FLARES, WING TIP
Magnesium lights electrically fired and used to facilitate landing at night.

FLATTENING OUT
A phrase used to describe the gradual decreasing of the gliding angle of a machine until it merges into the horizontal a few inches off the ground.

FLIGHT, A
An organization consisting of a small group of machines.

FLYING SPEED
The speed of a machine through the air necessary to maintain its support.

FORCED LANDING
See "Landing."

FORMATION FLYING
The practice of a group of machines keeping station in the air.

FUSELAGE
The body of a tractor machine.

GAP

The distance between the upper and lower wings of a biplane.

GAS BAG

Slang term for airships.

GLIDE

To descend with the engine cut off with the machine under control and at approximately the flying level speed.

GLIDING ANGLE

The angle that the fore-and-aft line of the machine makes with the horizon in order to make a correct gliding descent.

GROUND SPEED

The speed of the machine relative to the ground, which may be equal to, greater, or less than the air speed; therefore, ground speed is equal to air speed + or — wind speed.

HANGAR

An aeroplane shed.

HATE, TO COMMIT

To be extreme in doing a thing, i.e., excessive stunting near the ground.

HEAVY HANDED

Refers to a pilot who is clumsy with his controls and inclined to over-correct.

HEIGHT INDICATOR

See "Altimeter."

HOIK, TO

To make the machine climb steeply and suddenly.

HORIZON

The limit of ground in view.

HUN

Slang term for a person learning to fly.

INCIDENCE, ANGLE OF

The angle that the chord of a wing makes with the direction of motion relative to the air. A particularly muddling term, as it is often measured as the distance in inches that the front spar is above the rear spar when the machine is in the flying-level position.

INCLINOMETER

An instrument for showing the angle of the machine relative to the ground.

INDUCTION

The inlet stroke of the engine.

JOY STICK

See "Control Lever."

KATHEDRAL ANGLE

A machine is said to possess a kathedral angle when the wings slope downwards from the centre of the machine.

KEEL SURFACE

The side surface of a machine as opposed to the head-on surface.

KING POST

A bracing strut generally found on the top of controlling surfaces, such as rudder, ailerons and elevator, in which case it also acts as a lever.

KNOCK

A peculiar noise emanating from the engine and indicating some kind of mechanical trouble.

KNOT

A nautical mile per hour, i.e., it is wrong to speak of knots per hour.

LANDING

The action of a machine in coming to earth.

LANDING, FORCED

The action of a machine in coming to earth other than at the will of the pilot, i.e., in the case of the engine failing.

LEADING EDGE

The point or entering position of a wing.

LEEWARD

Away from the wind.

LEEWAY

Lateral drift to leeward.

LEFT AND RIGHT
> Always refer to the left and right of the machine and engine as seen by the pilot sitting in his seat.

LIFT
> The force exerted by the air on a plane in a direction perpendicular, or nearly so, to the motion.

LOG BOOK
> A book kept by pilots giving details of each flight.

LONGERONS
> The longitudinal members of the fuselage.

LOOP
> A manœuvre in which the machine, after flying straight, does an upward and backward turn or circle, and then continues in the same direction as before.

LUBBER LINE OR LUBBER'S POINT
> A mark on the body of a compass corresponding with the fore and aft line of the machine.

MACHINE
> The aeroplane as apart from the engine.

MOTOR
> An incorrect term for the power unit or engine.

M.P.H.
> Miles per hour.

NACELLE
> The body of a pusher machine.

NOSE
> The front part of a machine.

NOSE DIVE
> A very steep descent with or without engine.

NOSE HEAVY
> Backward pressure required on the control lever to make the machine fly level.

NOSE PIECE
> The front central portion of a rotary engine.

OUTRIGGER
> The framework connecting an elevator placed in front of the machine with the main planes.

PANCAKE
> To drop to earth from a height of a few feet owing to losing flying speed and flattening out too soon.

PEGGING DOWN
> Securing a machine by rope to pegs in the ground so as to prevent it capsizing in a wind.

PILOT
> A person controlling an aeroplane in the air.

PITCH OF A PROPELLER
> The forward distance that the propeller would travel if it were allowed to cut its way, without slip, through some medium such as butter.

PITOT TUBE
> Consists of two tubes, one open to the air flow and the other protected. The other ends of the two tubes are connected to the air-speed indicator. One tube is called the pressure tube and the other the suction, or static, tube.

PLANE
> Term used to apply to the supporting surfaces of a flying machine. The planes may be cambered, as in the case of the wings, or flat, as in the case of the tail plane.

PLANE, MAIN
> The wings of the machine.

POCKET AIR
> A disturbance in the air causing the machine to drop.

PROPELLER
> The airscrew driven by the engine which forces the machine through the air, generally known as the "Prop."

PROTRACTOR
> An instrument for measuring angles.

PUSHER
> A machine in which the propeller is fitted behind the main planes.

QUIRK
> A person learning to fly; slang term for pupil.

RACE, To

Refers to the practice of speeding up the revolutions of an engine to their maximum.

RADIAL

Refers to a type of engine in which the cylinders are set radially round the crankshaft, and are stationary.

RADIUS OF ACTION

The distance that a machine can fly from its starting point and return without replenishing the tanks. Greatly influenced by the wind factor.

REMOU

A disturbance in the air.

REMORQUE

A motor trailer for carrying aeroplanes.

REVS.

Short for revolution.

REVVING

Short for revolving.

R.F.

Representative fraction. A term indicating the scale of a map.

RIB, COMPRESSION

A rib designed to act as a strut between the front and rear spars of a wing.

RIBS

The members used in a wing to give strength and shape in a fore and aft direction. Often called "former ribs."

RIGHT AND LEFT

Always refer to the right and left of the machine and engine as seen by the pilot sitting in his seat.

RIGHT AND LEFT OF A MACHINE

Always refers to the machine as seen from the pilot's seat.

ROLL

A manœuvre in which the machine does a sideways turn or circle, and then continues in the same direction as before,

STRAINER

See "Turnbuckle." The word can also refer to gauge or chamois leather used to strain petrol through before refilling the tanks.

STREAMLINE

A shape of a body that offers the least resistance to its path through the air.

STRUTS

Vertical members uniting spars in upper and lower planes.

STUNTS

Unusual or exaggerated evolutions in the air.

"S" TURNS

A series of steeply-banked right and left-hand gliding turns (with engine off).

SWITCH

A device for allowing or interrupting the passage of electric current generally to the sparking plugs.

TACHOMETER

Engine revolution counter.

TAIL

A group of planes set behind the main planes and consisting of both vertical and horizontal surfaces, used to control the balance of the machine.

TAIL BOOM

The long spar connecting the main plane with the tail on a pusher machine.

TAIL HEAVY

Forward pressure required on the control lever to make the machine fly level.

TAIL PLANE

A fixed plane fitted parallel with the main plane, to which the elevator is attached.

TAIL PLANE, LIFTING

A fixed plane parallel with the main plane to which the elevator is attached. It also carries some of the weight of the machine.

Tail Skid
> See "Skid."

Taxying
> The progress of a machine on the ground with the engine running, though not fast enough to give flying speed.

Tee
> A ground sign indicating the direction of the wind. Originally an arrow, but it was found to be an improvement to widen the tip of the arrow until it became a T. The wind blows from the cross piece down the body of the T.

Tender
> A light motor lorry.

Throttle Lever
> Controls the amount of explosive mixture entering the engine.

Ticket, To Take
> To pass an elementary flying test, and thus be registered as a certified aviator.

Torque, Engine
> The reaction of a propeller which tends to cause the machine to turn about its longitudinal axis in a direction opposite to that in which the propeller is revolving.

Tractor
> A machine in which the propeller is fitted in front of the main planes.

Trailing Edge
> The rear edge of the wing.

Trestle
> Wooden frames or scaffolds designed to support the tail or wings of a machine when repairs are being carried out.

Triplane
> An aeroplane with three pairs of wings, set one above the other.

TRUEING UP
> Adjusting the rigging of a machine so as to correct its balance in the air.

TURNBUCKLE
> A fitting used to adjust the tension of wires to which it is attached. Also called a strainer.

UNDERCARRIAGE OR UNDERCHASSIS
> That part of a machine which carries the weight of the aeroplane on the ground, and also takes the shock of landing.

VEER, TO
> A change of wind in a clockwise direction, i.e., from N. to E.

VERTICLE BANK
> A loosely-applied phrase referring to any bank over 45 degrees.

VERY'S LIGHT
> A coloured light fired as a signal from a special form of pistol.

VOLPLANE
> A glide.

WARP
> The yarn running lengthwise in aeroplane fabric.

WARP, TO
> To move the control lever sideways so as to increase or decrease the incidence on a wing with a view to raising or lowering it.

WASH IN
> An increasing angle of incidence of a wing towards its wing tip.

WASH OUT
> A decreasing angle of incidence of a wing towards its wing tip.

WEFT
> The yarn running crosswise in aeroplane fabric.

WINDSCREEN
> A transparent screen mounted in front of the pilot and passenger to shield them from the rush of air by the machine in motion.

WIND SPEED
> The speed of the wind.

WIND UP
> To be frightened of going into the air.

WING
> The main supporting surface of an aeroplane.

WING TIP
> The right or left-hand extremity of a wing.

WING TIP SKIDS
> Semi-circular pieces of bamboo placed under the wing tips to take the shock off the wings, should the machine heel over on the ground.

WIRE, COMPENSATING OR BALANCING
> The wire connecting opposite ailerons of top or bottom planes.

WIRES, CONTROL
> See "Control Wires."

WIRES, DRIFT
> Used to transmit the head resistance set up by the wings to the main body of the machine.

WIRES, FLYING DRIFT
> Internal bracing wires of a wing connected from the front spar to the rear spar diagonally outwards in each cellule.

WIRES, FLYING OR LIFT
> Used to transmit the weight of the machine to the wings. They lie upwards and outwards.

WIRES, LANDING
> Used to take the weight of the wings when the machine is on the ground. They lie downwards and outwards.

WIRES, LANDING DRIFT
> Internal bracing wires of a wing connected from the front spar to the rear spar diagonally towards each cellule.

WIRE, SNAKE
> Fine wire twisted round other wires to prevent the latter fouling the propeller, should they break.

WIRES, WARP

Wires used to warp the ends of the wings to control the machine laterally.

ZOOM

To ascend very steeply after flying level at full speed.

Ingram Content Group UK Ltd.
Milton Keynes UK
UKHW022324050623
422929UK00005B/392